ECOLOGICAL EFFECTS OF ELECTRICITY GENERATION, STORAGE AND USE

CABI – who we are and what we do

This book is published by **CABI**, an international not-for-profit organisation that improves people's lives worldwide by providing information and applying scientific expertise to solve problems in agriculture and the environment.

CABI is also a global publisher producing key scientific publications, including world renowned databases, as well as compendia, books, ebooks and full text electronic resources. We publish content in a wide range of subject areas including: agriculture and crop science / animal and veterinary sciences / ecology and conservation / environmental science / horticulture and plant sciences / human health, food science and nutrition / international development / leisure and tourism.

The profits from CABI's publishing activities enable us to work with farming communities around the world, supporting them as they battle with poor soil, invasive species and pests and diseases, to improve their livelihoods and help provide food for an ever growing population.

CABI is an international intergovernmental organisation, and we gratefully acknowledge the core financial support from our member countries (and lead agencies) including:

Discover more

To read more about CABI's work, please visit: **www.cabi.org**

Browse our books at: **www.cabi.org/bookshop**,
or explore our online products at: **www.cabi.org/publishing-products**

Interested in writing for CABI? Find our author guidelines here:
www.cabi.org/publishing-products/information-for-authors/

ECOLOGICAL EFFECTS OF ELECTRICITY GENERATION, STORAGE AND USE

Peter A. Henderson

CABI is a trading name of CAB International

CABI	CABI
Nosworthy Way	745 Atlantic Avenue
Wallingford	8th Floor
Oxfordshire OX10 8DE	Boston, MA 02111
UK	USA
Tel: +44 (0)1491 832111	Tel: +1 (617)682-9015
Fax: +44 (0)1491 833508	E-mail: cabi-nao@cabi.org
E-mail: info@cabi.org	
Website: www.cabi.org	

A catalogue record for this book is available from the British Library, London, UK.

Library of Congress Cataloging-in-Publication Data

Names: Henderson, Peter A. (Peter Alan), author.
Title: Ecological effects of electricity generation, storage and use / Peter A Henderson, BSc PhD.
Description: Boston, MA, USA : CABI, [2018] | Includes bibliographical references and index.
Identifiers: LCCN 2018001822 (print) | LCCN 2018012085 (ebook) | ISBN 9781786392022 (ePDF) | ISBN 9781786392039 (ePub) | ISBN 9781786392015 (pbk : alk. paper)
Subjects: LCSH: Electric power production--Environmental aspects. | Renewable energy sources. | Electric power plants. | Environmental protection.
Classification: LCC TD195.E4 (ebook) | LCC TD195.E4 H48 2018 (print) | DDC 333.79/14--dc23
LC record available at https://lccn.loc.gov/2018001822

ISBN-13: 978 1 78639 201 5 (Pbk)
        978 1 78639 202 2 (PDF)
        978 1 78639 203 9 (ePub)

Commissioning editor: Ward Cooper
Editorial assistant: Emma McCann
Production editor: Tim Kapp

Typeset by SPi, Pondicherry, India
Printed and bound in the UK by Bell & Bain Ltd, Glasgow, G46 7UQ

Contents

Preface   vii

1. Our Need for Electricity and the Main Energy Sources Available   1

2. Hydroelectric Generation   8

3. Tidal Generation   19

4. Wave Power and Ocean Thermal Energy Conversion   28

5. Steam Turbines and Their Cooling Systems   36

6. Nuclear Generation   88

7. Coal and Oil-Fired Power Plants   94

8. Gas-Fired Power Plants   119

9. Wind Turbines and the Effects of Offshore Piling   129

10. Solar Power   154

11. Fuel Cells and Flow Batteries   169

12. Batteries   174

13. Biofuels and Waste-Powered Generation   180

14. Small-Scale and Mobile Electric Generators   189

15. Ecological Issues Relating to Transmission Lines   192

16. Geothermal Generation   199

17. Minimizing Environmental Damage While
     Generating Electricity Cost-Effectively   202

Index   219

Preface

The availability and consumption of electricity has been rising throughout the world for the last 100 years and will continue to rise for the foreseeable future. Affordable electricity transforms lives by making many tasks easier and less laborious and also by offering many opportunities for leisure and entertainment. To allow everyone on our planet to enjoy these benefits fully, we must ensure that the environmental impacts associated with electrical power generation are minimized. If we do not, we will degrade our world faster than the poor can acquire the technological benefits on offer.

The environmental impacts associated with generating electricity are far from trivial and include habitat degradation, species extinctions, accumulation of toxins in food species and atmospheric pollution sufficient to reduce life expectancy. We need to appreciate the full range of potential impacts of all the available technological options and not select for mass use any technologies without careful screening of all the merits and costs.

I have worked on the ecological impacts of power generation for about 40 years and have found that recently there has been a tendency to favour certain technologies because they reduce some particular class of impact. For example, solar, wind and nuclear generation have been particularly favoured, and in some countries heavily subsidized, because they are viewed as a way of reducing our CO_2 emissions and therefore reducing global warming. I have been at meetings where it has sometimes seemed that global warming is the only environmental issue of concern. It is unwise to take such a simple approach. Is it really sensible to kill millions of fish and crustaceans every year in a once-through cooling water system on a nuclear power station because a more environmentally protective closed-cycle system would fractionally reduce plant efficiency which might mean the release of more CO_2 in replacement generation? Does it make sense to build wind turbines because they are a low carbon generating method if they kill birds and bats? Some seem to believe so without making the calculations. They have stopped doing their own thinking. Some do not even want to admit the environmental problems linked to their favoured method of generation.

The key point I want the reader to acquire from this book is that all methods of electrical generation, storage and use have environmental consequences. The only truly environmentally positive approach is to embrace and seek out enhanced efficiency of generation and consumption. Fortunately, we are rapidly developing more efficient technologies and the next 20 years will produce many exciting possibilities. However, we must be cautious, tread lightly on the earth, and consider critically and carefully the environmental risks. Novel technologies inevitably produce surprising and unanticipated effects as the natural world responds to their deployment.

I would like to thank the helpful comments and ideas I have been given by my colleagues at Pisces, Richard Seaby and Robin Somes, and my friend at Oxford, the ever-active and thoughtful ecologist Clive Hambler.

Peter A. Henderson
Director Pisces Conservation Ltd and Senior Research Associate,
University of Oxford
August 2017

1

Our Need for Electricity and the Main Energy Sources Available

This book is about the environmental issues that need to be addressed when considering the options for the generation of electricity. It is important to be clear that the availability of a reliable and affordable supply of electricity is a fundamental requirement for everyday life in the present world. Our objective must therefore be to generate, supply and consume electricity in the most environmentally protective and cost-effective manner possible. Only by minimizing the environmental impact can we hope to supply sustainably plentiful electricity to our growing populations and allow the poorer people on our planet to enjoy the benefits presently enjoyed in wealthier countries. Good environmental practice must consider cost, because we must find cost-effective solutions if the poor are not to be excluded from the benefits of access to plentiful electricity. Some argue for environmentally protective approaches that increase costs without carefully considering the effect of increased costs, which inevitably affect the poor most. The availability of electricity, like all energy sources, is limited, and we cannot simply argue for ever higher production because this will inevitably increase environmental impacts. A theme that runs through this book is that all electricity-generating methods have their environmental impacts. It is therefore essential that we are efficient in our use of electricity and avoid waste if everyone is to benefit. Further, the best generating method will vary with location, so it is important to keep an open mind and consider the merits of the full range of alternatives. This is not to suggest there are no truly poor choices; there are, and they are often made. Governments are always attracted to large prestigious projects, often for the simple reason that they want to make a difference quickly and it is far easier to plan a single huge project than many smaller less impressive ones. There is always an engineering pressure group or union campaigning for the jobs the new giant nuclear plant, wind farm or dam will generate. They will have experts who will demonstrate the environmental benefits of the scheme. This book will have been of benefit if it helps in assessing the validity of the arguments of the industry experts.

It is worth reviewing the great benefits gained from access to electricity so that policies that seek environmental gain by restricting universal access to electricity or heavily restricting per capita use are rejected. However, this is not to suggest that everyone has a right to be profligate in their consumption of electricity. The data presented below suggest that, above a certain level of consumption, no great additional benefit in quality of life is gained from further consumption. Having four televisions on standby and consuming power, along with numerous phone and iPad chargers plugged in, as occurs in many US households, is simply silly and wasteful. It is claimed that approximately 75% of the electricity used in most American homes is consumed while the appliance is turned off. Idling devices certainly take a great deal of electrical power; the average desktop computer uses about 80 watts of electricity while not in use. Much can be gained if everyone views such behaviour as the private equivalent of throwing litter in the public park.

The Advantages of Electricity

Providing the costs are not prohibitive, electricity has the following advantages over other power sources.

- Convenience of conversion: it can be converted easily into heat, light and motion.
- Ease of control: it is easy to start, control the power output and stop electrical devices.
- Ease of transmission: unlike other energy sources such as coal and oil, which require massive transportation systems, electricity is exported easily via cables.
- Cleanliness: there is little pollution at the point of use from smoke, gases, etc. It is important to remember that the lack of pollution is at the point of use and it may not be the case at the point of production.

In addition, for small-scale power use, it can be a cheap energy source.

The Link Between Electricity and Standard of Living

For most of history, energy consumption has risen with living standards. Figure 1.1 shows the total energy production in various regions of the world. Three clear features are: (i) the general rising trend in all regions other than North America and Europe; (ii) the rapid recent increase in China; and (iii) the historically high energy production of the USA, which reflects the high standard of living.

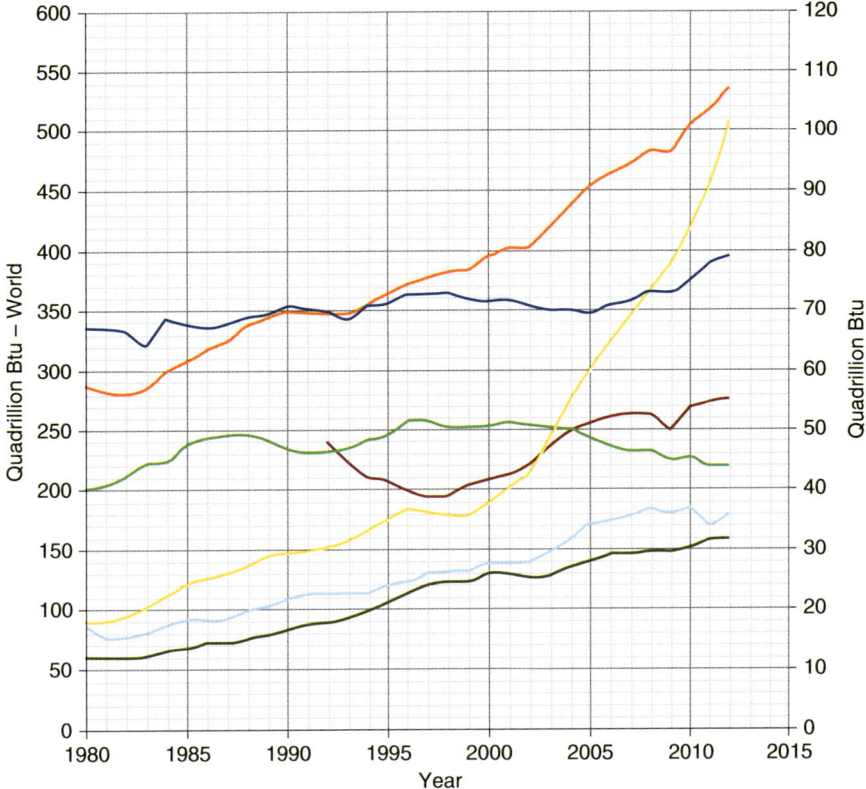

Fig. 1.1. World total primary energy production in quadrillion Btu. Note: World total on left y-axis, while regional figures are shown on right y-axis (approx. figures for 2010 and 2011, respectively). Red, world; dark blue, USA; yellow, China; maroon, Russia; green, Europe; light blue, Africa; black, South and Central America. (From US Energy Information Administration, public domain.)

One measure of the human condition is the Human Development Index (HDI),[1] a summary measure in terms of: (i) a long and healthy life; (ii) education; and (iii) the standard of living. The health dimension is assessed by life expectancy at birth. The education dimension is measured by mean years of schooling for adults aged 25 years and older, and expected years of schooling for children of school-entering age. The standard of living dimension is measured by gross national income per capita. The scores for the three HDI dimension indices are aggregated into a composite index using geometric mean. Figure 1.2 shows the relationship between the HDI and electricity consumption. The average world per capita consumption in 2004 was about 2490 kWh/person/year. It is clear that HDI rises rapidly as electricity consumption increases from zero to about 3000 kWh/person/year. Above this level of consumption, there is little improvement in the HDI.

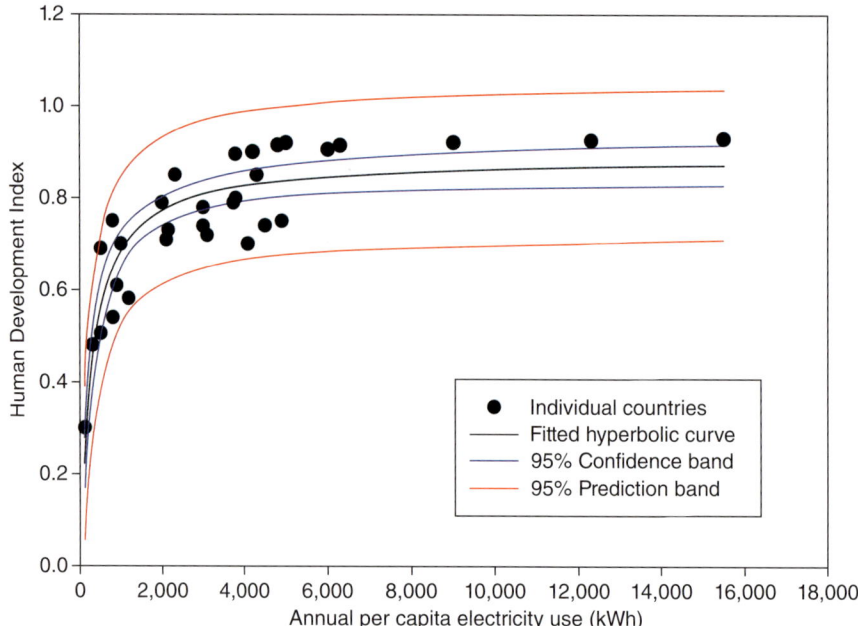

Fig. 1.2. The relationship between the HDI and per capita electricity consumption for the year 2003–2004. A hyperbolic curve has been fitted by regression showing that above a consumption of about 3000 kWh there is negligible gain in the HDI. (From UNDP, 2016.[1])

The Main Sources of Energy

Figure 1.3 shows the recent world trends in the main energy sources and their projected use. This figure has some interesting features. First, the carbon-based fuels, oil, coal and gas, have maintained their dominant position and are projected to remain so for the foreseeable future. Second, coal has shown the greatest change in relative contribution. In the early 2000s the energy contribution of coal increased dramatically. Nuclear generation has shown the lowest growth between 1990 and 2010 and there is no belief that it will undergo a dramatic increase in the projected future. Figure 1.3 makes clear that while there is considerable talk about the need to reduce CO_2 emissions from coal, oil and gas combustion, there is no belief this will occur. Indeed, for the foreseeable future coal will be a major source of world energy and the environmental impacts of coal-based electricity generation will be an important issue.

The seemingly smooth increases in the various fuel types consumed at a world level is not reflected in all regions. Possibly a better reflection of what we can expect in the future is shown by recent developments in the USA. Figure 1.4 shows long-term energy consumption in the USA. There is a tendency for the use of different energy sources to rise and fall

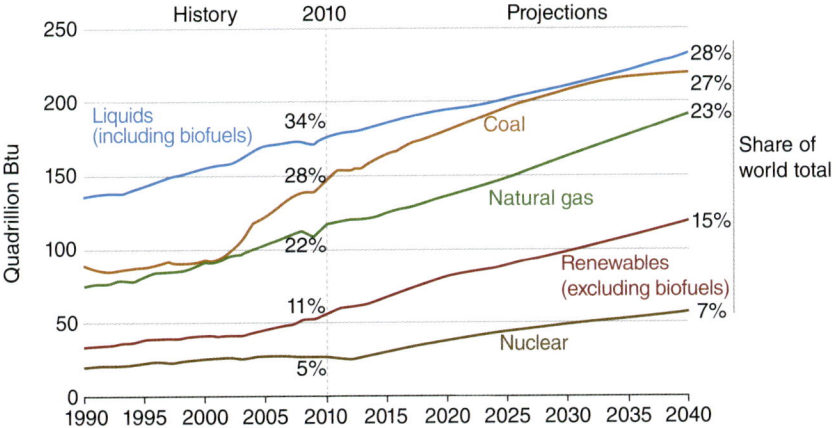

Fig. 1.3. World energy consumption outlook from the International Energy Outlook. (US Energy Information Administration, International Energy Outlook 2013; via presentation by Adam Sieminski, IEO2013, public domain.)

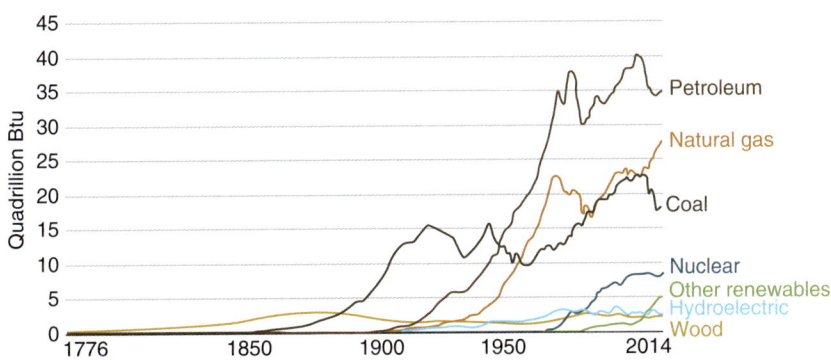

Fig. 1.4. Energy consumption of the USA, 1776–2014. (US Energy Information Administration, public domain.)

over time. At the beginning of the 19th century, wood was the dominant fuel, and interestingly the energy contributed by wood is still at a similar level. The 19th century was characterized by the rise in coal consumption, which continued to about 1950, when petroleum and natural gas started their dramatic increase in importance. Coal had a second resurgence from the 1960s onwards, but after about 2000 it went into a steep decline, as did petroleum. Hydroelectric generation was essentially fully developed by the mid-20th century and has stopped growing. Nuclear had a dramatic rise from the early 1960s, but now appears to be at a plateau and about to decline. The new wave of growth is now in other renewables, and it seems inevitable that their contribution will continue to rise for some time. Unlike hydroelectric generation, they are not limited by suitable river locations for dams.

The Growth in Electricity Consumption

Figure 1.5 shows the recent growth in world electricity generation. It is notable that fossil fuel generation is dominant and the relative proportion of renewable generation has been gradually increasing since the 1980s. This hides very different patterns in different regions. Figure 1.6 shows the

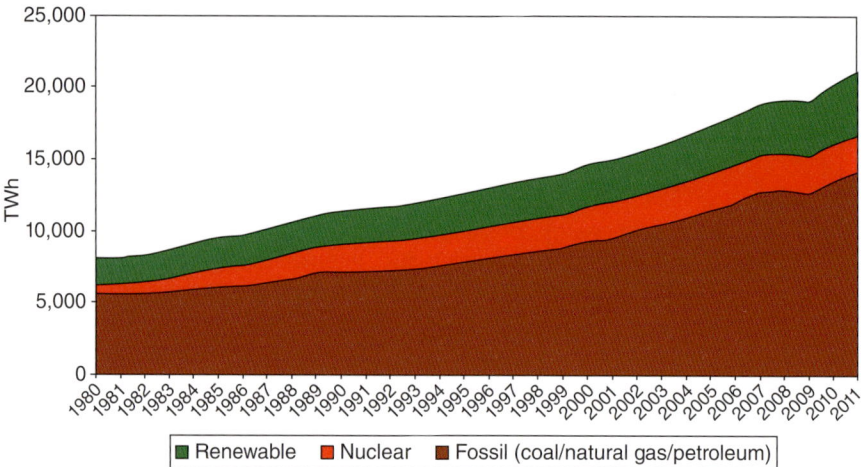

Fig. 1.5. Annual electricity net generation in the world, 1980–2011. (From US Energy Information Administration, public domain.)

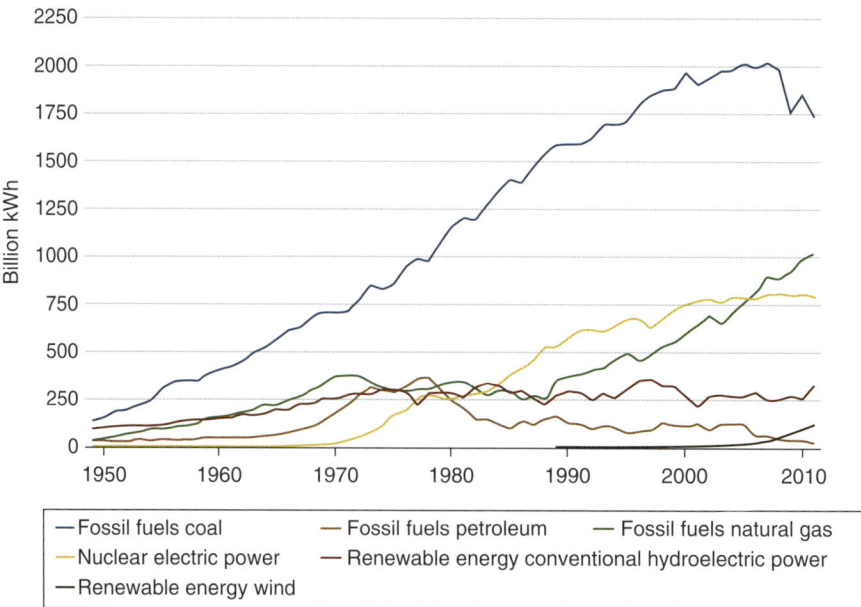

Fig. 1.6. US electricity generation mix by fuel type, 1949–2011. (From US Energy Information Administration, public domain.)

electricity generation mix for the USA from 1949 until 2011 (more recent data are difficult to obtain). Energy use in the USA is doubling about every 20 years. The first point to note is the dominance of the fossil fuels coal and gas. Second, while coal consumption is now in clear decline, it is to a great extent being replaced by natural gas and not solar and wind. In 2014, US natural gas production was the highest ever recorded. For all of the 21st century to date, nuclear power generation has not increased, and given the lack of new build and closure of older plant, it is inevitable that a decline in nuclear generation will soon be apparent. This is important at a global level because the USA presently generates more electricity using nuclear power than any other country. While an enormous expansion in renewable energy production in the USA is now underway, the US Energy Information Administration believes the country will still be using non-renewable energy sources for most of its energy needs in 2040.

There are now signs that the increased efficiency of modern appliances and lighting is beginning to reduce electricity demand in some western countries. One of the most dramatic changes has been in domestic lighting: LED lights use 75% less energy and last 25 times longer than the old incandescent bulbs. The use of microwave cookers has also reduced the energy used in heating food. In the UK, for example, electricity demand plateaued and then fell over the period 2003–2012. This seems to be linked to a combination of energy efficiency gains, high energy prices and economic recession.[2]

Notes

[1] United Nations Development Programme (UNDP) (2016) *Human Development Index (HDI)*. Available at: http://hdr.undp.org/en/content/human-development-index-hdi (accessed September 2017).

[2] GOV.UK (2017) *Historical electricity data: 1920 to 2016*. Available at: https://www.gov.uk/government/statistical-data-sets/historical-electricity-data-1920-to-2011 (accessed September 2017).

2   Hydroelectric Generation

Water has long been used as a source of energy – water wheels were in use over 2000 years ago. Hydroelectric power converts the potential (occasionally kinetic) energy of water via turbines to electricity. The most common type of hydroelectric power plant uses a dam on a river to store water in a reservoir. Water released from the reservoir flows through a turbine, spinning it, which in turn activates a generator to produce electricity. Hydroelectric power can be generated without a large dam; some hydroelectric power plants channel water along a canal and through a turbine rather like old water mills. Some of the largest and most impressive engineering projects undertaken by man are for hydroelectric generation. The Three Gorges Dam (Fig. 2.1) spans the Yangtze River, in the P.R. China, and has the largest installed capacity of 22,500 MW. In 2014 the Three Gorges Dam held the record for total electricity generated of 98.8 TWh, but in 2016 this was surpassed by the Itaipú Dam on the Brazil/Paraguay border (Fig. 2.2) when it set a new world record of 103.1 TWh.

Pumped storage plants are hydroelectric generators designed to store energy. When excess energy is available in the supply grid, water is pumped to an upper reservoir. At times of peak demand, this water is then run through turbines to generate electricity. Pumped storage plants have the advantage that they can supply electricity rapidly when required. An interesting example of this type of plant is Dinorwig Power Station, Wales. The main plant is installed in an immense cavern within a mountain (Fig. 2.3). Water is pumped up to an upper reservoir and released when electricity is required. Dinorwig comprises six 300-MW GEC generator/motors coupled to Francis-type reversible turbines and operates at about 75% efficiency.

There are also many installed small and micro hydroelectric power stations, often generating 1–20 MW (Fig. 2.4). There has been a considerable

Fig. 2.1. The Three Gorges Dam on the Yangtze River, China. (Le Grand Portage Derivative work: Rehman, published under a CC BY 2.0 license via Wikimedia Commons.)

Fig. 2.2. Panoramic view of the Itaipú hydroelectric dam from the Brazilian side. (Photo courtesy of Martin St-Amant, published under a CC BY-SA 3.0 license via Wikimedia Commons.)

increase in small-scale hydroelectric installed capacity since 2000. In places with a suitable flowing water source, domestic-scale hydroelectric generation is often used. In Europe, many old water mills have been converted to small-scale electricity generation.

Installed Capacity

Hydroelectric generation presently contributes about 16.6% of global electricity demand and represents 70% of all renewable electricity generation.[1] There is a small growth in production estimated to be about 3.1% per year for the next 25 years. As shown in Table 2.1, China has the greatest installed capacity and generated 1064 TWh in 2014 (16.9% of consumption). Hydropower is the dominant renewable energy resource in the USA, producing 6.5% of total electricity consumed. Norway is unique in generating almost all of its electricity by hydropower – a particularly notable fact given their large oil and gas resources.

Fig. 2.3. The Dinorwig power station in Wales. (Photo courtesy of Denis Egan, published under a CC BY 2.0 license via Wikimedia Commons.)

Fig. 2.4. The 2-MW Lochay hydroelectric power station, Perthshire, Scotland. (Photo courtesy of Dr Richard Murray, published under a CC BY 2.0 license via Wikimedia Commons.)

Table 2.1. The ten countries with the largest hydroelectric generating capacity in 2014. (Data from *Renewables 2016 Global Status Report*, http://www.ren21.net/wp-content/uploads/2016/06/GSR_2016_Full_Report_REN21.pdf[1])

Country	Annual hydroelectric production (TWh)	Installed capacity (GW)	Capacity factor	% of total production
China	1064	311	0.37	18.7
Canada	383	76	0.59	58.3
Brazil	373	89	0.56	63.2
USA	282	102	0.42	6.5
Russia	177	51	0.42	16.7
India	132	40	0.43	10.2
Norway	129	31	0.49	96.0
Japan	87	50	0.37	8.4
Venezuela	87	15	0.67	68.3
France	69	25	0.46	12.2

Ecological Impacts of Hydroelectric Generation

The cost of hydroelectricity is relatively low, making it a competitive source of renewable electricity. Once a hydroelectric complex is constructed, the project produces no direct waste and has a considerably lower output level of greenhouse gases than fossil fuel powered energy plants. Hydropower is generally considered to be a clean, renewable source of energy, emitting a very low level of greenhouse gases when compared with fossil fuels, with low operating costs once constructed. It is shown below that this may not always be the case; they can cause the release of mercury compounds leading to toxin accumulation in top predators and perhaps, most surprisingly, the release of appreciable amounts of greenhouse gases.

Land use and habitat destruction

One of the most contentious issues relating to hydroelectric generation is the amount of land lost to the reservoir. This can vary considerably, depending on the size of the project and the topography of the land. Dams placed across rivers in shallow valleys will create extensive shallow reservoirs.

The worst example of excessive land use is the Balbina hydroelectric plant, which was built in the Amazon forest, an area notable for the lack of variation in the elevation of the land. This single dam flooded 2360 km[2] of forest for 250 MW power-generating capacity (>2000 acres per MW).[2] The initial flooding of the reservoir resulted in extensive forest destruction because the trees died (see Fig. 2.5). The dam was built to give the city of Manaus a reliable electricity supply. One consequence was that the Waimiri-Atroari indigenous people were displaced from their homeland.

Fig. 2.5. View over the reservoir showing killed trees at the Balbina Dam (Usina Hidreletrica de Balbina), in Amazon, Brazil. (Photo courtesy of Seabirds, published under a CC BY-SA 3.0 license via Wikimedia Commons.)

It has been claimed that because of the methane released from the reservoir, the Balbina Dam emits more greenhouse gases than the equivalent level of generation using coal as an energy source. Such projects are extraordinarily destructive of habitat.

As another extreme example, the reservoir of the Three Gorges Dam in China holds 39.3 km³ of water with a total surface area of 1045 km²; the reservoir flooded a total area of 632 km² (see Fig. 2.6). Such a huge land take resulted in extensive impact in terms of lost wildlife habitat, agricultural land, villages and towns and scenic vistas.

As an average, Fthenakis and Kim (2009)[3] estimated the typical land use of hydroelectric plants as, on average, 4000 m²/GWh; only biomass power plants consume more land (for comparative figures see Chapter 17, this volume).

Water use

It is often claimed that a hydroelectric station consumes no water once constructed, unlike coal or gas plants. This is incorrect. Macknick *et al.* (2011)[4] reviewed operational water consumption for electricity generating technologies and concluded that hydroelectric plants in the USA consumed between 1425 and 18,000 gallons of water per megawatt-hour

November 7, 2006

April 17, 1987

Fig. 2.6. The Yangtze River in the vicinity of the Three Gorges Dam (lower right). Landsat 7 acquired the top image on 7 November 2006, after the main wall was complete. Landsat 5 acquired the bottom image on 17 April 1987. The lower image has been recoloured to more closely match the colours of the upper one. (Photo courtesy of Sagredo at English Wikipedia, public domain via Wikimedia Commons.)

(gal/MWh) of electricity generated. This consumption was through evaporation losses from the reservoir surface. The median loss of 4491 gal/MWh is actually higher than the loss for generation technologies such as coal and gas plants that use cooling towers (see Chapter 5, this volume).

Wildlife impacts

Hydroelectric dams have huge impacts on aquatic ecosystems, both as a consequence of a change in the aquatic environment and also directly by interfering with migration and causing injury or death to animals that pass through the turbines.

Impacts of rapid changes in downstream flow rates

Cushman (1985)[5] reviewed the impacts of rapid changes in flow below hydroelectric facilities resulting from variable generation linked to peaks in demand. He concluded that short-term, recurring disturbances of aquatic systems below dams have important ecological consequences including reduced productivity in the tailwaters, and reduced abundance, diversity and productivity of riverine organisms, particularly fish.

Methylmercury bioaccumulation

In dammed reservoirs[6] methylmercury bioaccumulation occurs in fish and in the consequent consumption of fish by humans. Research in northern Canadian reservoirs has shown that methylmercury in fish can reach very high levels. For example, pike (*Esox lucius*) and walleye (*Stizostedion vitreum*) in La Grande Reservoir in the James Bay region of Québec reached approximately six times background levels, or more than seven times the Canadian marketing limit of 0.5 mg/g (Verdon *et al.*, 1991).[7] The link between newly flooded organic matter and methylmercury production obviously suggests remediation by removal, burning or covering of vegetation and soil organic matter before flooding to reduce the severity of the problem. However, this is impractical to carry out in large reservoirs. Active methylmercury production in newly built reservoirs following impoundment may persist for up to 10 years and remain in fish for 30 years.[8] Studies have shown that soil organic matter content and Hg concentration, the ratio of flooded area to total water surface area, water residence time and primary productivity in the biota system play an important role in net Hg methylation in reservoir systems. While methylmercury may directly harm fish populations, there is little evidence to demonstrate this is occurring; the main concern is the effect of consuming these fish on human populations.

Interference with fish migrations

Many riverine fish make extensive migrations between spawning and feeding grounds. Many rivers hold species that either migrate to sea to spawn (such as eels) or move up rivers to spawning sites (such as salmonids). Dams can cause populations of migratory fish to collapse. Population collapse has long been known to occur. For example, shads are members of the herring family that migrate into freshwater to breed. In the River Severn in England, Day (1890)[9] noted that by the late 19th century twaite shad had declined greatly in abundance. He clearly considered that the construction of weirs and dams was the major cause of this decline because of the loss of access to spawning sites. 'It passes up the Severn to the Teme, up which it ascends so far as the Powick weir. But it likewise (as in the case of Allis shad) is a fish that will at no distant date probably be almost or quite extinct in this river, for it is unable to obtain access to its spawning beds, and these forms of shad, which were of great moment to the fisheries of the lower Severn, are fallen to great poverty'.

Fish injury and mortality resulting from fish passage through hydroelectric plants, including turbines, spillways, sluiceways and other passage routes during downstream migration, are reported frequently. Injury and mortality are caused in a number of ways, including: abrasions, scrapes and mechanical strikes from turbine blades; injuries from turbulence and shear owing to different water velocity experienced across the body length; rapid change in water pressure; violent cavitation bubble collapse; and impact with debris screens and trash racks.

The impact of hydroelectric power generation has been studied in migrating European eels in the River Meuse in the Netherlands using radio-telemetry. Transponders implanted into silver eels enabled their movements to be monitored near the hydropower installations (Winter *et al.*, 2006).[10] This study demonstrated that migrating eels showed a reluctance to pass the turbines and subsequently suffered estimated turbine-related mortality of 16–26%.

The typical approach to mitigate against the impact of migratory barriers involves the construction of fish passes (see Fig. 2.7); the Sustainable

Fig. 2.7. A fish ladder constructed for the Pitlochry Dam, Scotland. (Photo courtesy of Pisces Conservation Ltd.)

Development Commission report provides a useful summary of the issues with mitigation (Sustainable Development Commission, 2007).[11]

At some dams fish lifts have been installed. For example, the Tuilières dam on the Dordogne, France, has a fish lift to help migrating sturgeon, salmon and shad ascend the dam. This small 18-MW facility, which was originally built in 1905, when it was the largest hydroelectric power station in Europe, has a visitor centre to watch the operation of the lift.

Changes in community structure

Changes in the physical environment brought about by dam closure have profound effects on the aquatic ecosystem. With large dam projects, the impact can be seen over huge geographical scales.

The Aswan High Dam, an embankment dam built across the Nile in Aswan, Egypt, between 1960 and 1970, gives good examples of the major changes that can occur (see Figs 2.8 and 2.9). The reservoir, named Lake Nasser (Fig. 2.9), is 550 km long and a maximum of 35 km wide, with a surface area of 5250 km.[2]

Some of the major ecological impacts following construction of the Aswan dam include:[12]

- Downstream phytoplankton density increased from 160 to 250 mg/l because reduced levels of silt in the water increased light and hence photosynthesis.
- Reduction in nutrient flux to the Mediterranean Sea.
- Decline in fisheries in brackish Delta lakes.

Fig. 2.8. View from the embankment of the Aswan dam down the River Nile. (Photo courtesy of P.A. Henderson.)

Fig. 2.9. Lake Nasser behind the Aswan Dam, Egypt. (Photo courtesy of P.A. Henderson.)

- The commercial fishery in the Nile Delta changed. While the number of fish species, fish abundance and average size declined at some locations, at others abundance and size increased.
- The sardine fishery in the eastern Mediterranean declined, probably because of water quality issues. The sardine catch off the Egyptian coast declined from 18,000 tons in 1962 to 460 tons in 1968, but then gradually recovered to 8590 tons in 1992.
- Shrimp catches declined after closure of the dam.
- Demersal fish catches declined after closure, but then partly rebounded, probably because of an increase in motorized boats in the decade after 1970.
- Accelerated migration of Red Sea fish into the Mediterranean which began with the opening of the Suez Canal but had been reduced by the flow of the Nile into the sea. The river had previously created conditions at the mouth of the canal that were unattractive to Red Sea fish.
- Water quality in the River Nile has declined and drinking water requires greater treatment.

Life-cycle global warming emissions from hydroelectric plants

Global warming emissions are produced during the installation and dismantling of hydroelectric power plants, but, surprisingly, it has been found that emissions of greenhouse gases during operation can also be significant.[13] Such emissions vary greatly, depending on the size of the reservoir and the nature of the land that was flooded by the reservoir.

Life-cycle emissions from large-scale hydroelectric plants built in semi-arid regions are not large at approximately 0.06 pounds of CO_2 equivalent per kWh. However, estimates for life-cycle global warming emissions from hydroelectric plants built in tropical areas or temperate peatlands are much higher. After the area is flooded, the vegetation and soil in these areas decomposes and releases both CO_2 and methane. The exact amount of emissions depends greatly on site-specific characteristics. However, current estimates suggest that life-cycle emissions can be over 0.5 pounds of CO_2 equivalent per kWh.[14] To put this into context, estimates of life-cycle global warming emissions for natural gas-generated electricity are between 0.6 and 2 pounds of CO_2 equivalent per kWh and estimates for coal-generated electricity are 1.4 and 3.6 pounds of CO_2 equivalent per kWh.[15]

Notes

[1] *Renewables 2016 Global Status Report* (2016) Available at: http://www.ren21.net/wp-content/uploads/2016/06/GSR_2016_Full_Report_REN21.pdf (accessed September 2017).

[2] Fearnside, P.M. (1989) Brazil's Balbina Dam: environment versus the legacy of the Pharaohs in Amazonia. *Environmental Management* 13(4), 401–423.

[3] Fthenakis, V. and Kim, H.C. (2009) Land use and electricity generation: a life-cycle analysis. *Renewable and Sustainable Energy Reviews* 13(6), 1465–1474.

[4] Macknick, J., Newmark, R., Heath, G. and Hallett, K.C. (2011) *Review of Operational Water Consumption and Withdrawal Factors for Electricity Generating Technologies (No. NREL/TP-6A20-50900).* National Renewable Energy Laboratory (NREL), Golden, Colorado.

[5] Cushman, R.M. (1985) Review of ecological effects of rapidly varying flows downstream from hydroelectric facilities. *North American Journal of Fisheries Management* 5(3A), 330–339.

[6] Rosenberg, D.M., Berkes, F., Bodaly, R.A., Hecky, R.E., Kelly, C.A. and Rudd, J.W. (1997) Large-scale impacts of hydroelectric development. *Environmental Reviews* 5(1), 27–54.

[7] Verdon, R., Brouard, D., Demers, C., Lalumiere, R., Laperle, M. and Schetagne, R. (1991) Mercury evolution (1978–1988) in fishes of the La Grande hydroelectric complex, Quebec, Canada. *Water, Air & Soil Pollution* 56(1), 405–417.

[8] Meng, B., Feng, X., Qiu, G., Li, Z., Yao, H., Shang, L. and Yan, H. (2016) The impacts of organic matter on the distribution and methylation of mercury in a hydroelectric reservoir in Wujiang River, Southwest China. *Environmental Toxicology and Chemistry* 35(1), 191–199.

[9] Day F. (1890) Notes on the fishes and fisheries of the Severn. *Proceedings of the Cotswold Naturalists Field Club* 9, 202–219.

[10] Winter, H.V., Jansen, H.M. and Bruijs, M.C.M. (2006) Assessing the impact of hydropower and fisheries on downstream migrating silver eel, *Anguilla anguilla*, by telemetry in the River Meuse. *Ecology of Freshwater Fish* 15, 221–228.

[11] Sustainable Development Commission (2007) *Turning the Tide. Tidal Power in the UK.* Sustainable Development Commission, London.

[12] Rosenberg, D.M., Berkes, F., Bodaly, R.A., Hecky, R.E., Kelly, C.A. and Rudd, J.W. (1997) Large-scale impacts of hydroelectric development. *Environmental Reviews* 5(1), 27–54.

[13] Ibid.

[14] National Academy of Sciences (2010) *Electricity from Renewable Resources: Status, Prospects, and Impediments.* The National Academies Press, Washington, DC. Available at: http://www.nap.edu/openbook.php?record_id=12619 (accessed September 2017).

[15] IPCC (2011) *IPCC Special Report on Renewable Energy Sources and Climate Change Mitigation.* IPCC, Geneva, Switzerland.

3 Tidal Generation

There are many designs for the conversion of tidal energy into electricity. These can be divided into two groups: (i) tidal stream generators that use tidal kinetic energy and (ii) tidal range generators that use the potential energy in the raised head of impounded waters. Suitable resources for tidal stream and tidal range generators are not necessarily found in the same localities.

While estimates of global potential may vary, it is widely agreed that tidal stream energy capacity could exceed 120 GW globally. The UK has one of the largest marine energy resources in the world, estimated to be more than 10 GW and representing about 50% of Europe's tidal energy capacity.

Tidal power is far from evenly distributed, and only selected coastal localities can ever offer viable locations for tidal generation. The distribution of spring tide tidal energy around the British coast is shown in Fig. 3.2. In Great Britain, the majority of tidal stream resource is in the north of Scotland and around the Pentland Firth, also around Alderney, Anglesey and the Strangford Lough area in Northern Ireland. However, the main tidal range resource lies in the western estuaries, particularly the Severn Estuary which has at least a 12-m tidal range in the vicinity of the proposed barrage (maximum about 15 m), which is the second or third largest tidal range in the world. Other large tidal ranges are found further north in the Mersey Estuary and the Solway.

Present installed capacity

Total global tidal power-generating capacity is tiny. South Korea has the greatest capacity at 511 MW, followed by France with 246 MW, the United Kingdom with 139 MW, Canada with 40 MW, Belgium with 20 MW, China with 12 MW and Sweden with 11 MW.

Fig. 3.1. Aerial view of the tidal barrage on the Rance and of Saint Malo, France. The first tidal power station constructed. (Photo courtesy of Tswgb, public domain via Wikimedia Commons.)

Tidal Barrages

A tidal barrage is a wall-like structure constructed across the width of an estuary or coastal inlet to produce electricity. It uses the potential energy held in the difference in height in water as the tides rise and fall (see Fig. 3.3). A differential water height is created by a dam and the resulting potential energy is captured via the flow passing across turbines. One of the largest tidal barrage schemes proposed is the Severn Barrage between England and Wales (Fig. 3.4), designed to hold 192 turbines with a combined maximum power output of 8640 MW during flow and 2000 MW average power over the tidal cycle. The 17 TWh of electricity it would generate per year corresponds to about 6% of UK consumption. For the Severn Barrage proposal, water would pass across the turbines at a velocity of 11.4 m/s (see Fig. 3.5).

There are few completed power-generating barrages. The first tidal power station was built at La Rance, France in the 1960s and has an installed capacity of 240 MW (see Fig. 3.1). Sihwa Lake Tidal Power Station in South Korea is presently the largest tidal power station with a total capacity of 254 MW. The tidal barrage uses a seawall constructed in 1994 for flood defence and agriculture. Ten 25.4-MW submerged bulb turbines are used to generate on tidal inflows only. The working basin area is 30 km² but may be reduced by land reclamation and the creation of freshwater lagoons. The first tidal power site in North America is the Annapolis Royal Generating Station, Nova Scotia, which opened in 1984 on an inlet

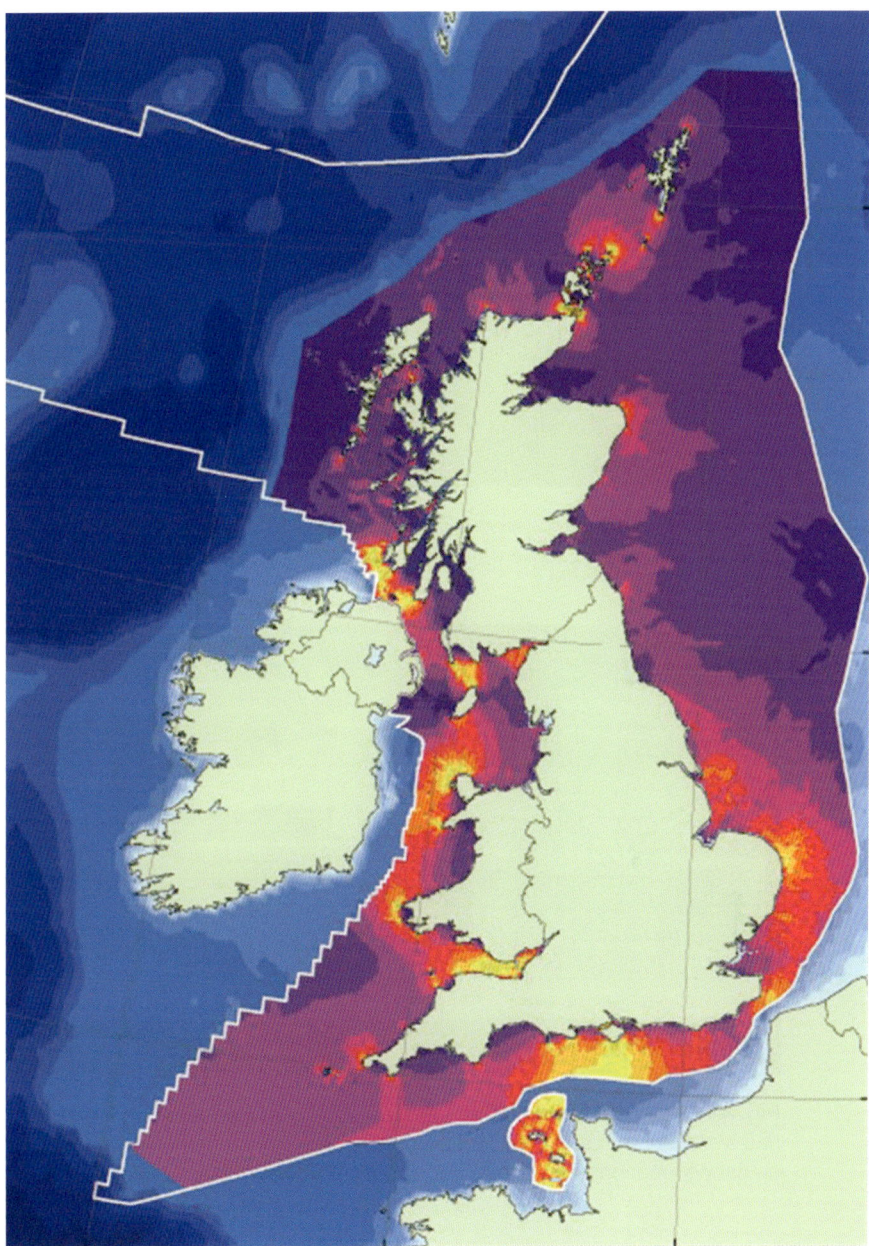

Fig. 3.2. The UK marine energy resource: mean spring tidal power (yellow shows higher power regions). (From *Atlas of Marine Energy Resources*, Crown Copyright.)

of the Bay of Fundy and has an installed capacity of 20 MW. The Bay of Fundy is famous as having the greatest tidal range in the world. The Jiangxia Tidal Power Station, south of Hangzhou in China, has been operational since 1985, with an installed capacity of only 3.2 MW at present.

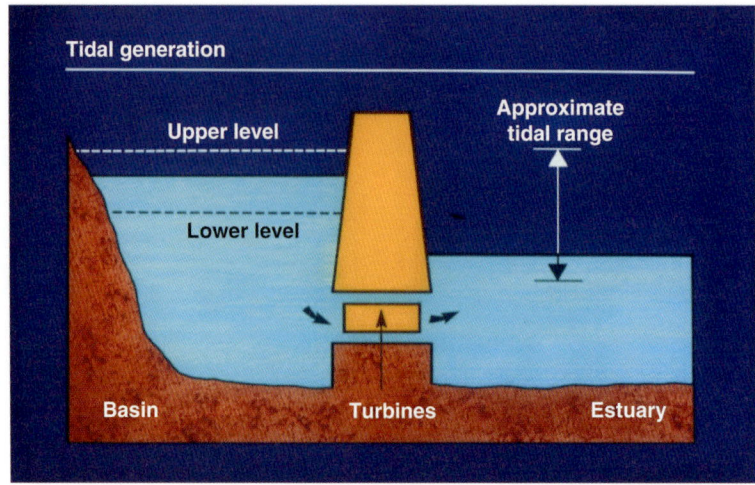

Fig. 3.3. Diagrammatic cross-section of a tidal barrage showing the position of the turbines at the base of the dam. (From Pisces Conservation Ltd.)

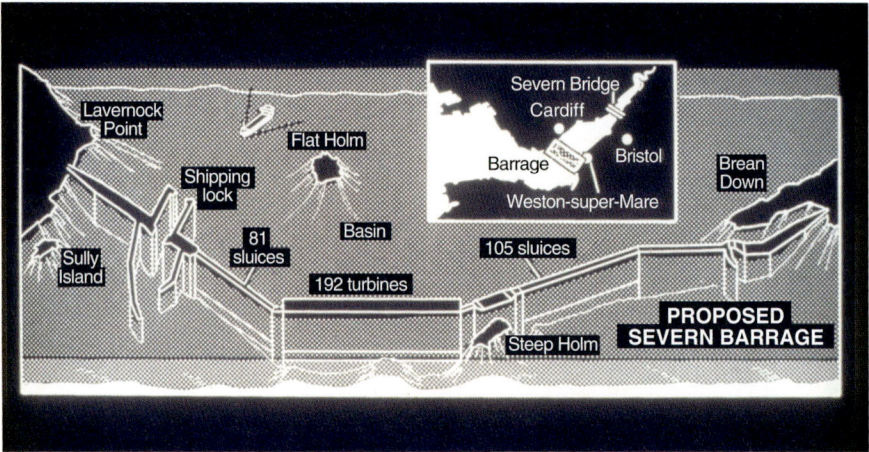

Fig. 3.4. The basic design of the proposed Severn Barrage. The Bristol Channel has a maximum tidal range of more than 14 m making it one of the most suitable estuarine locations for a major tidal barrage. (From Pisces Conservation Ltd.)

Ecological impacts of barrages

The installed capacity of power-generating barrages is small at present and therefore care needs to be taken when assuming current plants reflect the full potential environmental issues.

Barrages have the potential to radically alter the functioning of estuarine ecosystems. One key area of concern is the impact on marine life. The following sections summarize the other concerns.

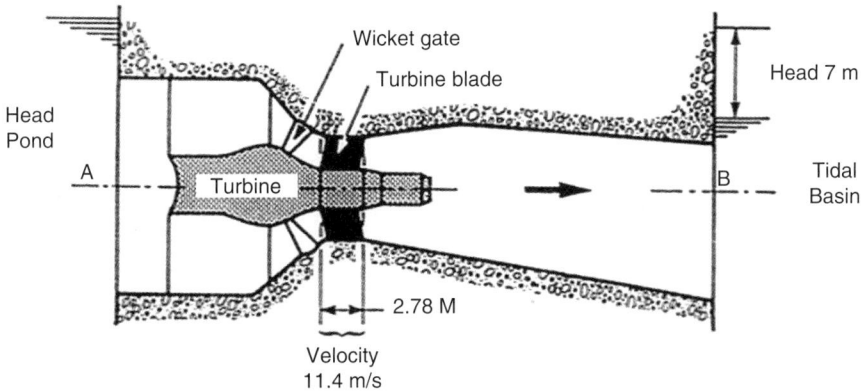

Fig. 3.5. A diagrammatic cross-section of the turbine design proposed for the Severn Barrage in the United Kingdom. (From Pisces Conservation Ltd.)

Physical changes
By impounding the water for part of the tide there will inevitably be some changes in the estuary basin. The following list, derived in part from Wolf *et al.* (2009),[1] covers the main areas of concern.

- Local scouring around the outflow regions of the turbines and sluices, and siltation of the basin because the tidal and residual flows are modified.
- Seawater vertical stratification is reduced as the tidal flows are reduced.
- Suspended solids decline, causing increased light penetration.
- Stratification of waters of different density increases because the reduction in mixing causes differences in salinity and temperature.
- Reduced saline water penetration.
- Contaminant build-up caused by reduced flushing.
- Re-suspension of contaminated sediments in areas of increased flow.
- Land drainage affected by the increase in average water level within the basin.
- A change in the area of intertidal habitat within the basin.

Changes in primary production
Increased nutrient concentrations combined with increased light produce increased primary production. Changes in primary productivity have both good and bad consequences. While they can lead to increased animal biomass, they can also cause mass mortalities if oxygen is consumed by the plankton during the night.

Species migrating across the barrage line and turbine passage injury
Many studies have considered in some detail how a barrage might affect anadromous and catadromous fish, such as eel, salmon, sea trout and shad. However, there are considerably more migratory species living in a typical estuary than is often realized. Many marine fish undertake seasonal

migrations up and down estuaries. Further, many invertebrates also migrate; for example, shrimps, prawns and crabs.

For fish, the single most important problem arising from tidal power generation is mortality of migratory fish during passage through a turbine. The type of fatal injury that has been observed when a fish is hit by a turbine blade is shown in Fig. 3.6. Fish can be mutilated and killed without striking a solid surface. Figure 3.7 shows the corpse of a fish with the head ripped off at the gills after contact with a powerful shear force caused by hitting a jet of water moving at right angles to the orientation of the fish. Fish injury and death was identified as a significant problem when the barrage proposals were considered (Department of Energy, 1989)[2] and has been highlighted again in the SDC review (Sustainable Development Commission, 2007a and 2007b).[3, 4] A study to identify factors causing injury to fish during turbine passage concluded that there was insufficient understanding of the specific injury mechanisms (Solomon, 1988).[5] Another report on the impact of the tidal barrage in the Rance basin on the northern coast of Brittany concluded that there is little evidence that

Fig. 3.6. The remains of a fish struck by a turbine blade retrieved downstream of the turbine showing the type of catastrophic injury that can be inflicted. (Photo courtesy of Pisces Conservation Ltd.)

Fig. 3.7. The remains of a fish in which the head has been snapped off at the gills following exposure to a powerful current while passing through a generating turbine. (Photo courtesy of Pisces Conservation Ltd.)

Table 3.1. Predicted injury rates for fish passing through a Severn barrage turbine. (From Sustainable Development Commission, 2007a.[3])

Fish (length)	Injury rate (%)
Adult salmon (1000 mm)	40
Salmon smolt (150 mm)	10
Adult European eel (700 mm)	28
Juvenile shad (70 mm)	53

turbine-related deaths have been a significant problem. At this locality, migratory fish and cephalopods apparently pass through the turbines, or via sluice gates, unharmed (Retiere, 1994).[6] In contrast, other evidence suggests that the La Rance barrage has had an impact on trout migration (SAGE, 2004).[7]

Fish passage studies conducted at the Annapolis Royal low-head tidal turbine in the Bay of Fundy in Canada also suggest that turbine-related mortality can be significant (Dadswell and Rulifson 1994).[8] At this site, passing though the turbine was estimated to result in 20–80% mortality per passage. The mortality rate depended on the fish species involved, fish size and the efficiency of the turbine operation.

More recently, additional work on the injury caused to fish by turbines has been simulated for various stress factors. Mortality due to rapid pressure change, hydrologic shear stress and turbulence, cavitation and runner blade strike were each modelled separately (Turnpenny *et al.*, 2000)[9] and the results have led to greatly improved estimates of fish injury for key migratory species (Table 3.1).

IMPACT ON BIRDS. The potential effect of tidal barrages on marine bird populations and waders in particular has been a major concern for barrage proposers. Clark (2006)[10] discusses the potential issues and highlights three main impacts:

- Changes in size and nature of the intertidal areas within an estuary.
- Effects on saltmarshes.
- Displacement of birds at closure.

Clark makes the point that estuaries differ in their utility to birds and therefore the impact, in terms of birds affected per unit of generation, can vary greatly between estuaries.

Tidal Stream Generators

A number of small tidal stream generators are now operating. About 2 km off the north-east tip of Scotland, the MeyGen project is currently the largest tidal stream project in the world. The MeyGen has permits for up to 398 MW of tidal stream capacity to be installed, but the present grid capacity is limited to 252 MW.

The SeaGen project, Northern Ireland

The world's first commercial-scale tidal turbine was placed in Strangford Lough, Northern Ireland, in July 2008 (Fig. 3.8). The 1.2 MW SeaGen project was developed by Marine Current Turbines (MCT). Total power output by September 2012 was 5 GWh. The SeaGen tidal turbine is a free stream tidal energy device converting tidal flow energy into electricity. It comprises twin 16-m diameter rotors connected to a generator via a gearbox.

Environmental issues relating to tidal stream generation

During the construction phase, the main issues relate to drilling, piling and cabling creating noise and disturbance. For a discussion of the impact of underwater noise, see Chapter 9 (this volume). The direct loss of habitat is not likely to be significant as the footprint of tidal flow generators is expected to be small.

Operational impacts potentially include: changes in sediment erosion, transportation and deposition; underwater noise; and the risk of collision,

Fig. 3.8. SeaGen is the world's first commercial tidal generator in Strangford Lough, Northern Ireland. (Photo courtesy of Fundy, published under a CC BY-SA 3.0 license via Wikimedia Commons Creative Commons.)

both above and under water. Monitoring work since construction concluded that porpoise activity declined during installation, but no long-term changes in abundance of either seals or porpoises have been attributed to the presence or operation of the device.[11] No major changes in bird numbers in the vicinity of the device were reported.

Tidal Lagoons

A variant on the tidal barrage is to build tidal lagoons. These are circular retaining walls embedded with turbines that can capture the potential energy of tides. The lagoons could also act as pumped storage schemes (see Chapter 2, this volume). The proposed Tidal Lagoon in Swansea Bay, Wales, UK, is at the most advanced stage of planning and could be the first in the world.

Ecological issues related to tidal lagoons are comparable with those for tidal barrages. The problems of fish injury during passage through the turbines are the same as in a conventional tidal barrage design (see above).

Notes

[1] Wolf, J., Walkington, I.A., Holt, J. and Burrows, R. (2009) Environmental impacts of tidal power schemes. *Proceedings of the Institution of Civil Engineers – Maritime Engineering* 162(4), 165–177.

[2] Department of Energy and Climate Change (2009) *Severn Tidal Power*. Available at: http://webarchive.nationalarchives.gov.uk/20110405184751/http://www.decc.gov.uk/EN/Default.aspx?n1=3&n2=51&n3=58&n4=60&n5=171 (accessed 5 April 2018).

[3] Sustainable Development Commission (2007a) *Turning the Tide. Tidal Power in the UK.* Sustainable Development Commission, London.

[4] Sustainable Development Commission (2007b) *Tidal Power in the UK. Research Report 3 – Review of Severn Barrage Proposals.* Sustainable Development Commission, London.

[5] Solomon, D.J. (1988) *Fish Passage Through Tidal Energy Barrages.* Support Unit, Harwell, UK. Contractors Report No. ETSU TID4056.

[6] Retiere, C. (1994) Tidal power and the aquatic environment of La Rance. *Biological Journal of the Linnean Society* 51, 25–36.

[7] SAGE (2004) Schéma d'aménagement et de Gestion del'Eau de la Rance et du Frémur. Arrêté préfectoral du 5 avril 2004. Tome 1 du SAGE Frémur Baie de Beaussais – La Rance et la Frémur en 2002 – tat des lieux.

[8] Dadswell, M.J. and Rulifson, R.A. (1994) Macrotidal estuaries – a region of collision between migratory marine animals and tidal power development. *Biological Journal of the Linnean Society* 51, 93–113.

[9] Turnpenny, A.W.H., Clough, S., Hanson, K.P., Ramsay, R. and McEwan, D. (2000) Risk assessments for fish passage through small, low-head turbines. Contractor's Report No. ETSU H/06/00054/REP. Fawley Aquatic Research Laboratories Ltd, Report to the Energy Technology Support Unit (ETSU), Harwell, Oxfordshire, UK.

[10] Clark, N.A. (2006) Tidal barrages and birds. *Ibis* 148(s1), 152–157.

[11] Royal Haskoning (2011) *SeaGen Environmental Monitoring Programme Final Report.* Available at: https://tethys.pnnl.gov/publications/seagen-environmental-monitoring-programme-final-report (accessed March 2018).

4 Wave Power and Ocean Thermal Energy Conversion

Using the power of waves to generate electricity is an old idea which has never achieved large-scale commercial viability. Because there has been renewed recent interest stimulated by climate warming concerns and the desire for low-carbon electricity generation, a number of test facilities have been established recently. The European Marine Energy Centre (EMEC), the first marine energy test facility, was established in 2003 in Orkney, Scotland, and has deployed more wave and tidal energy devices than at any other single site in the world (see http://www.emec.org.uk/). At present, North America has shown little interest in wave energy generation. In the USA, there are three experimental sites that each has a single device: Makah Bay, Washington; Kaneohe Bay, Hawaii; and off the coast of New Jersey. Uihlein and Magagna (2016)[1] review the current state of wave power-generating technologies.

Wave power resources are not evenly distributed. The map of wave power around the United Kingdom (Fig. 4.1) shows the greatest energy is available off the Scottish west coast and far out into the Atlantic Ocean.

Types of Wave Generator

There is a wide range of potential designs, the most common of which are summarized below (see Fig. 4.2).

Point absorber buoys

Point absorber buoys float on the surface of the water, anchored by cables to the seabed. The rise and fall of the swell drives hydraulic pumps to generate electricity.

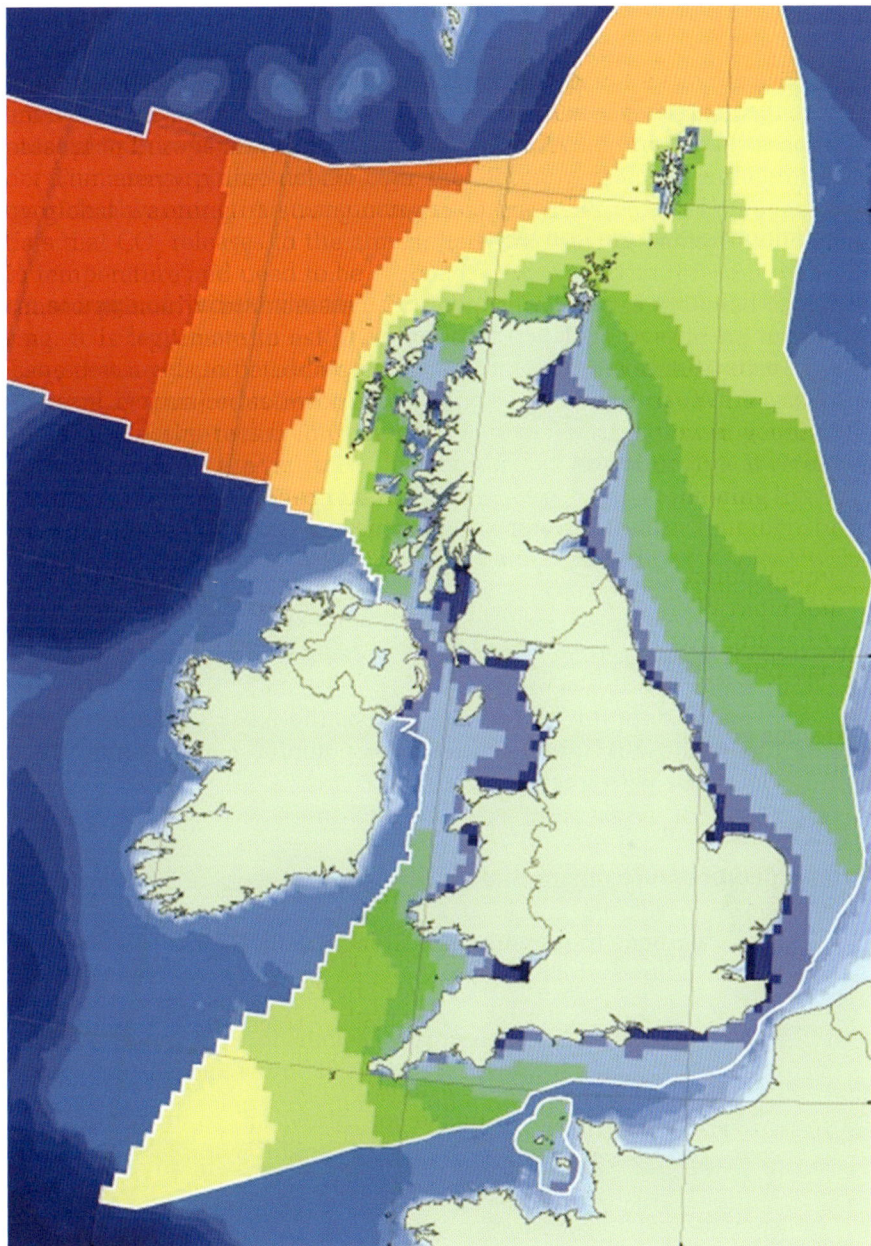

Fig. 4.1. Annual mean wave power around the British Isles. Red shows the region with the highest energy waves. (From *Atlas of Marine Energy Resources*, Crown Copyright.)

Surface attenuators

Surface attenuators are made up of multiple connected floating segments oriented perpendicular to the waves. The flexing motion generated by the swell drives a hydraulic pump. An example is the Pelamis machine

(Fig. 4.3), which is composed of a series of semi-submerged cylindrical sections linked by hinged joints. As waves pass along the machine, the induced motion is used by hydraulic cylinders to pump high pressure oil through hydraulic motors which drive electrical generators. Electricity is fed via cable to the sea bed and passed ashore via an undersea cable.

Oscillating wave surge converter

Oscillating wave surge converters have a fixed part attached to a structure or the seabed and a part that is moved by the swell and this motion is converted to electricity.

Oscillating water column

In an oscillating water column the swell compresses air in a chamber, which passes into an air turbine to generate electricity.

Overtopping device

In an overtopping device, waves raise the water level in a chamber above that of the surrounding water. This potential energy is then converted to electricity with low-head turbines.

Submerged pressure differential

The difference in pressure at different locations below a wave are used to induce a flow, which drives a turbine and electrical generator.

The four most commonly proposed machines are point absorber buoys, surface attenuators, oscillating water columns and overtopping devices.

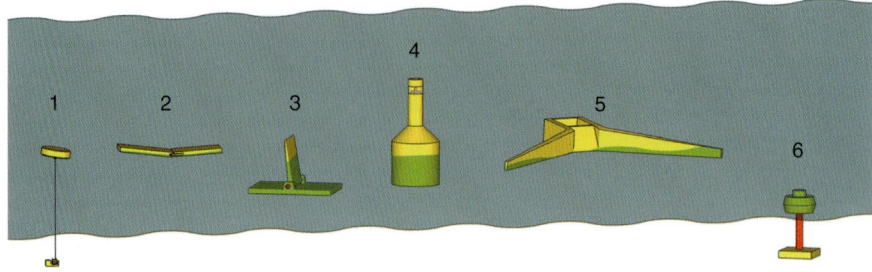

Fig. 4.2. Generic wave energy concepts: 1, point absorber; 2, attenuator; 3, oscillating wave surge converter; 4, oscillating water column; 5, overtopping device; 6, submerged pressure differential. (From Ingvald Straume, public domain via Wikimedia Commons.)

Fig. 4.3. Pelamis Wave Energy Converter on site at the European Marine Energy Test Centre (EMEC). (Photo courtesy of P123, public domain via Wikimedia Commons.)

Ecological Impacts

The ecological implications of large-scale wave energy generation cannot be assessed fully because there are no large-scale wave power generation facilities in operation. Given our present lack of research knowledge and operational experience, it is best to simply list some areas that will need to be considered in any ecological assessment. Ecological issues are discussed in Lewis *et al.* (2011).[2] It is important to note that, as with other renewable, low carbon, electrical generation technologies, there has been a lack of long-term in-depth research into the ecological consequences of large-scale development.

- Large-scale generation would result in alterations to the physical environment and habitat change. For example, energy removed from the waves may affect the shoreline and the distribution of sediments. Numerical modelling has predicted that wave generation will extract between 3% and 15% of incident wave energy,[3] and this energy reduction is likely to affect wave shoaling, sediment transport, beach building and coastal erosion. Seabed and midwater habitats will experience changes in currents, mixing within the water column and sedimentation.
- Like offshore wind farms, there are ecological issues relating to the installation and siting of undersea cables and anchors (see p. 136). They may create artificial reefs.
- The electromagnetic fields generated by electrical transmission cables may interfere with fish and other marine life that are sensitive to

electrical fields. This problem also occurs with wind farm cables, so experience gained from their operation will be useful (see p. 136).

- The noise generated may cause avoidance or flight reactions in marine life. There is, at present, little information on the level of undersea noise generated.
- The presence of the buoys and other surface structures may affect the distribution of fish and marine mammals. Birds are likely to be attracted to structures as resting sites while hunting. Pelagic fish may aggregate around structures.
- The size of wave installations offshore have the potential to affect commercial fishing. It is estimated that 1000 linear wave converters would occupy an area of about 1 km^2.
- Animals may become entangled in lines and cables.
- Large-scale structures may interfere with or alter migratory routes of marine animals such as salmon, diving mammals and cephalopods (e.g. squid). Some marine organisms use the earth's magnetic field for navigation and may become disorientated by the magnetic field around transmission cables.
- Cables need to be laid across the seabed or buried, creating construction disturbance. These cables also need to be brought onshore, resulting in disturbance to beaches (see wind turbine issues, p. 136).
- If generating units break free, they would likely cause damage to the seabed and shore.
- Pollutant release, for example hydraulic fluids, needs to be considered with some designs.
- Antifouling technologies will be required, some of which have ecological impacts.
- Corrosion protection, if required, may have ecological impacts.

Ocean Thermal Energy Conversion

Ocean thermal energy conversion (OTEC) uses the temperature differential between cold, deep, ocean waters and the much warmer surface waters. This technology requires the greatest possible temperature gradient and is therefore most viable in tropical waters where the temperature differential is greater than 20°C (see Fig. 4.4).

Pelc and Fujita (2002)[4] review both the possible designs for OTEC plants and their environmental effects. OTEC is categorized as open cycle, closed cycle or hybrid cycle, depending on the type of working fluid used by the plant. Open-cycle OTEC (see Fig. 4.5) uses seawater as the working fluid, which is flash-evaporated in a vacuum chamber to produce steam to drive the turbine. Closed-cycle OTEC (see Fig. 4.6) uses ammonia or another low boiling-point fluid, which is heated by the surface water, vaporized and then used to drive the turbine and generator. Hybrids of these two approaches have also been proposed. OTEC plants can be built both onshore and offshore.

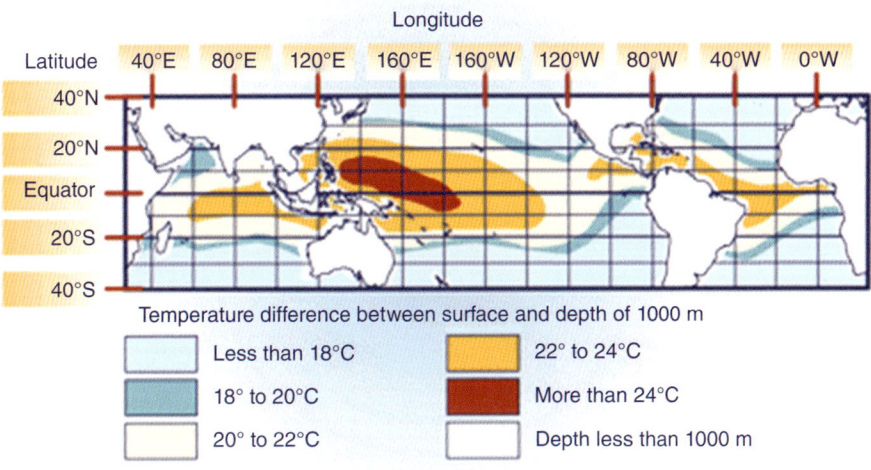

Fig. 4.4. Red and orange areas indicate parts of the ocean in which OTEC is a viable option due to the temperature gradient. (From StefKa81, public domain via Wikimedia Commons.)

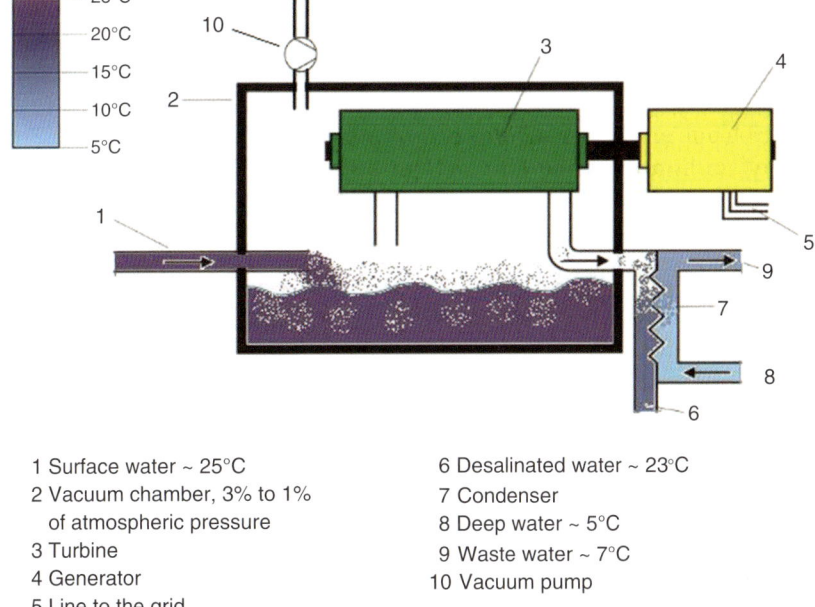

1 Surface water ~ 25°C
2 Vacuum chamber, 3% to 1%
  of atmospheric pressure
3 Turbine
4 Generator
5 Line to the grid

6 Desalinated water ~ 23°C
7 Condenser
8 Deep water ~ 5°C
9 Waste water ~ 7°C
10 Vacuum pump

Fig. 4.5. Diagram of an open-cycle OTEC plant. (From Lumos3, published under a CC BY-SA 1.0 license via Wikimedia Commons.)

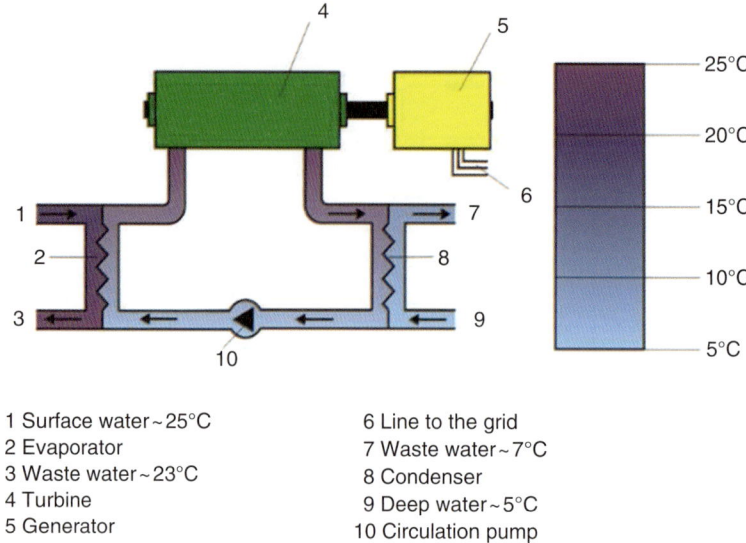

1 Surface water ~ 25°C   6 Line to the grid
2 Evaporator   7 Waste water ~ 7°C
3 Waste water ~ 23°C   8 Condenser
4 Turbine   9 Deep water ~ 5°C
5 Generator   10 Circulation pump

Fig. 4.6. Diagram of a closed-cycle OTEC plant. (From Lumos3, published under a CC BY-SA 1.0 license via Wikimedia Commons.)

Installed OTEC capacity

OTEC is still an experimental technology with no large-scale facilities in operation. A few small-scale experimental units are operating, including a 100-kW unit in Japan, a small unit in Hawaii and a larger unit in India producing about 1 MW. The Lockheed Martin Company has announced plans for a 10-MW offshore plant in China.

Ecological impacts of OTEC

As with wave generation, we do not have the operational experience of utility-scale OTEC plants to be able to assess ecological impacts. I therefore present a list of areas of concern that need to be addressed when assessing a proposal for a large-scale implementation.

- Large discharges of mixed warm and cold water would be released near the surface, creating a plume of cool, denser, sinking water.
- Thermal impacts on local habitats such as coral reefs may occur. The continual use of warm surface water and cold deep water may, over long periods of time, lead to slight warming at depth and cooling at the surface.
- Increased nutrient loading resulting from the discharge of upwelled water could produce adverse impacts on the naturally low nutrient ecosystems typical of tropical waters. For example, creating algal blooms.

- Toxic chemicals, such as ammonia and chlorine, might be discharged accidently following plant failure.
- Impingement of large organisms and entrainment of small organisms may occur. For floating OTEC plants, these would likely be weakly swimming organisms such as juvenile fish, jellyfish and invertebrates. A small amount of sequestered CO_2 is held in deep waters, which would be released to the atmosphere.
- Construction onshore and the construction of transmission cables to the shore will produce shoreline impacts.

Notes

[1] Uihlein, A. and Magagna, D. (2016) Wave and tidal current energy – a review of the current state of research beyond technology. *Renewable and Sustainable Energy Reviews* 58, 1070–1081.

[2] Lewis, A., Estefen, S., Huckerby, J., Musial, W., Pontes, T. and Martinez, J.T. (2011) Ocean energy. In: Edenhofer, O., Pichs-Madruga, R., Sokona, Y., Seyboth, K., Matschoss, P., Kadner, S., Zwickel, T., Eickemeier, P., Hansen, G., Schlömer, S. and von Stechow, C. (eds) *IPCC Special Report on Renewable Energy Sources and Climate Change Mitigation.* Cambridge University Press, Cambridge, UK, pp. 497–533.

[3] Largier, J., Behrens, D. and Robart, M. (2008) The potential impact of WEC development on nearshore and shoreline environments through a reduction in nearshore wave energy. In: Nelson, P.A., Behrens, D., Castle, J., Crawford, G., Gaddam, R.N., Hackett, S.C. *et al.* (eds) *Developing Wave Energy in Coastal California: Potential Socio-Economic and Environmental Effects.* California Energy Commission, PIER Energy-Related Environmental Research Program and California Ocean Protection Council, Sacramento, California, USA.

[4] Pelc, R. and Fujita, R.M. (2002) Renewable energy from the ocean. *Marine Policy* 26(6), 471–479.

5 Steam Turbines and Their Cooling Systems

This chapter reviews ecological issues common to all power-generating systems that use steam turbines, irrespective of their power source. About 86% of total electrical output is generated using steam turbines. Nuclear, coal, oil, gas, solar boilers and biofuel power sources generate using steam turbines and share common environmental impacts relating to the need for cooling to condense steam.

When possible, power stations are placed close to water sources able to supply condenser cooling water. The water requirement per unit of electricity generated varies with the cooling system design and power source. It is common, even in the scientific literature, to read that large cooling water flows are needed to cool the reactor or generators; this is a misleading statement and so it is useful to first summarize why power stations equipped with steam turbines need large cooling systems.

Steam from the boiler is passed through a series of steam turbines to generate the turning motion that generates the electricity. At the exhaust end of this turbine set, the steam is condensed to form liquid water, which is then passed back into the boiler to become steam again. Water reduces greatly in the volume occupied as it condenses, forming almost vacuum conditions in the condenser. It is the formation of this very low pressure that continually draws the steam through the turbine. The cooling system is used to condense the exhaust steam and carry away the heat. This waste heat is an inevitable thermodynamic cost inherent in the use of a heat engine. It is true that large power stations also have other pieces of machinery that need to be cooled, but the cooling of electrical components, gear boxes and bearings is small when compared with the need to remove heat from the condensers.

Power stations may use one of a number of cooling options which are described briefly in the next section.

Fig. 5.1. Maintenance of a low-pressure section of a steam turbine at the Balakovo Nuclear Power Plant. (Photo courtesy of Alexander Seetenky, published under a CC BY-SA license.)

Cooling Methods

The classification of cooling systems can be confusing. The following three binary distinctions are in general use:

- Dry air-cooled compared with evaporative wet-cooled. In evaporative cooling, water is lost to the atmosphere and heat is removed by latent and sensible heat transfer. With dry cooling, water is not lost to the atmosphere and only sensible heat is transferred.
- Open compared with closed systems. In an open system, the coolant (water) is in contact with the environment; in a closed system, the coolant circulates through the plant without contact with the environment.
- Direct compared with indirect systems. In a direct system, there is one heat exchanger where the coolant and the medium to be cooled exchange heat. In an indirect system, there are at least two heat exchangers and a closed secondary cooling circle between the process or product to be cooled and the primary coolant.

Once-through cooling systems

Direct, once-through, cooling systems
In direct once-through systems, water is pumped from a source (e.g. a river, lake, sea or estuary) via large water inlet channels directly to the plant.

After passing through heat exchangers or condensers, the heated water is discharged directly back into the surface water (Figs 5.2 and 5.3). The heat is transferred from the turbine steam water to the coolant through the wall of the condenser tubes.

POTENTIAL ENVIRONMENTAL ISSUES.   For once-through systems, the major environmental issues are:

- Those associated with the use of large amounts of water, such as impingement and entrainment of fish and other aquatic life (see p. 45).
- The discharge of heated water (see p. 76).
- Sensitivity to biofouling and the need to add antifouling agents (see p. 82).
- Corrosion and scaling problems.
- The release of heavy metals (see p. 109).
- The use of additives and the resulting emissions to water (e.g. sawdust or ferrosulfate).
- Construction of intake structures, intake canals, etc.
- Changes in water flow and bed scour.
- If fish return systems are fitted, the attraction of predatory fish and birds to the discharge, which alters the local ecology.

Once-through cooling systems with cooling tower

At some localities, once-through systems are combined with a cooling tower to cool the discharge before it is released to the receiving surface water (Fig. 5.4). This is done in situations where cooling water may recirculate and raise the temperature of the cooling water intake of the same plant, or where there is a serious risk of the heated effluent damaging sensitive environments. This configuration has been used at both coastal and inland power stations.

POTENTIAL ENVIRONMENTAL ISSUES.   For once-through systems with cooling towers, the major environmental issues are:

- Those associated with the use of large amounts of water, such as impingement and entrainment of fish and other aquatic life. This is lower than for a simple once-through system.
- The discharge of heated water.
- Sensitivity to biofouling and the need to add antifouling agents.
- Corrosion and scaling problems.
- The release of heavy metals.
- The use of additives and the resulting emissions to water.
- Construction of intake structures, intake canals, etc.
- Changes in water flow and bed scour.
- Cooling towers in saline water systems can cause salt spray drift.
- Possible build-up of pathogens in the cooling towers. Controlled by the use of biocides.

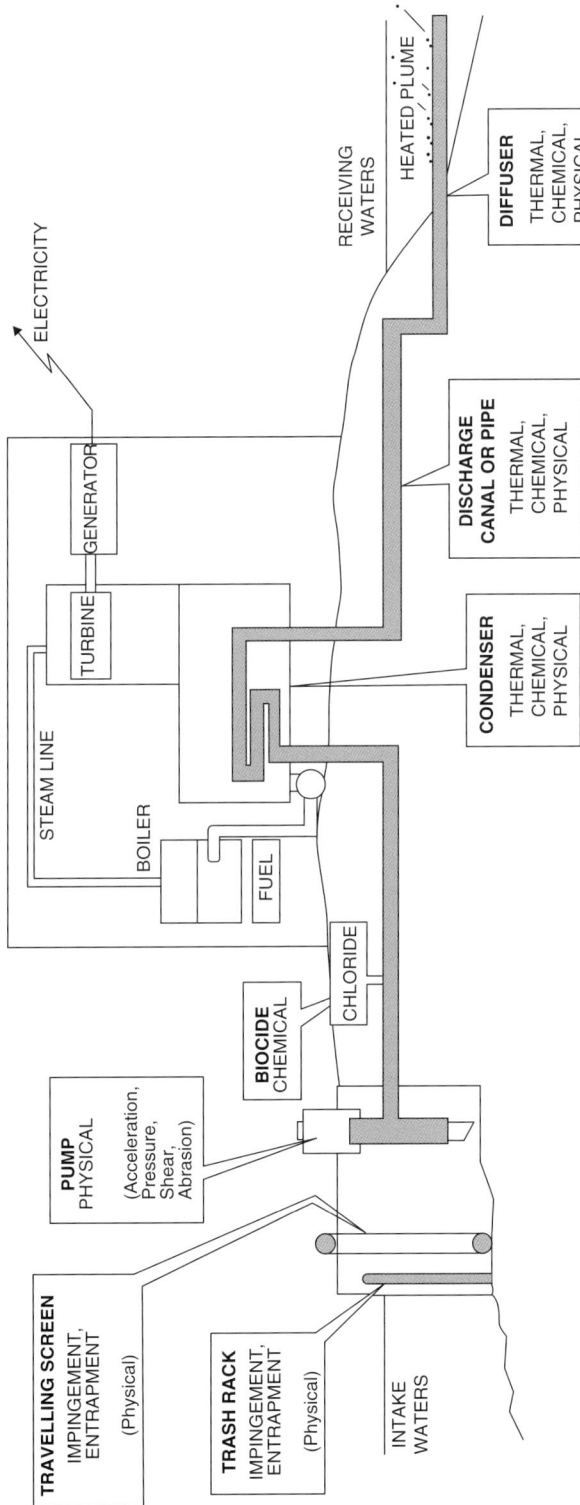

Fig. 5.2. Schematic representation of a direct once-through cooling system. (From Pisces Conservation Ltd.)

Fig. 5.3. The discharge of Hinkley Point Nuclear Power Station, Somerset, UK, showing the cooling water discharge in the foreground. This nuclear power station has a direct, once-through cooling system. (Photo courtesy of Pisces Conservation Ltd.)

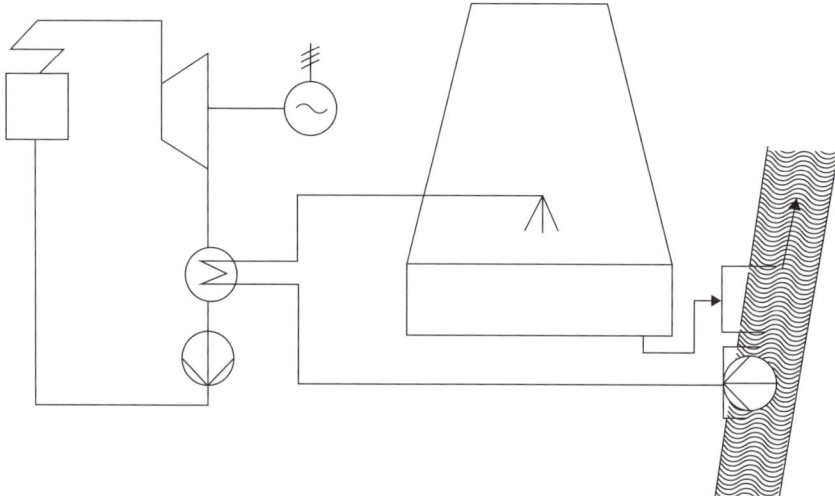

Fig. 5.4. Schematic representation of a direct once-through cooling system with a cooling tower as applied in the power industry. (From Pisces Conservation Ltd.)

Open recirculating cooling systems

Open recirculating cooling systems are also referred to as open evaporative cooling systems (Fig. 5.5). In these systems, cooling water that is led through

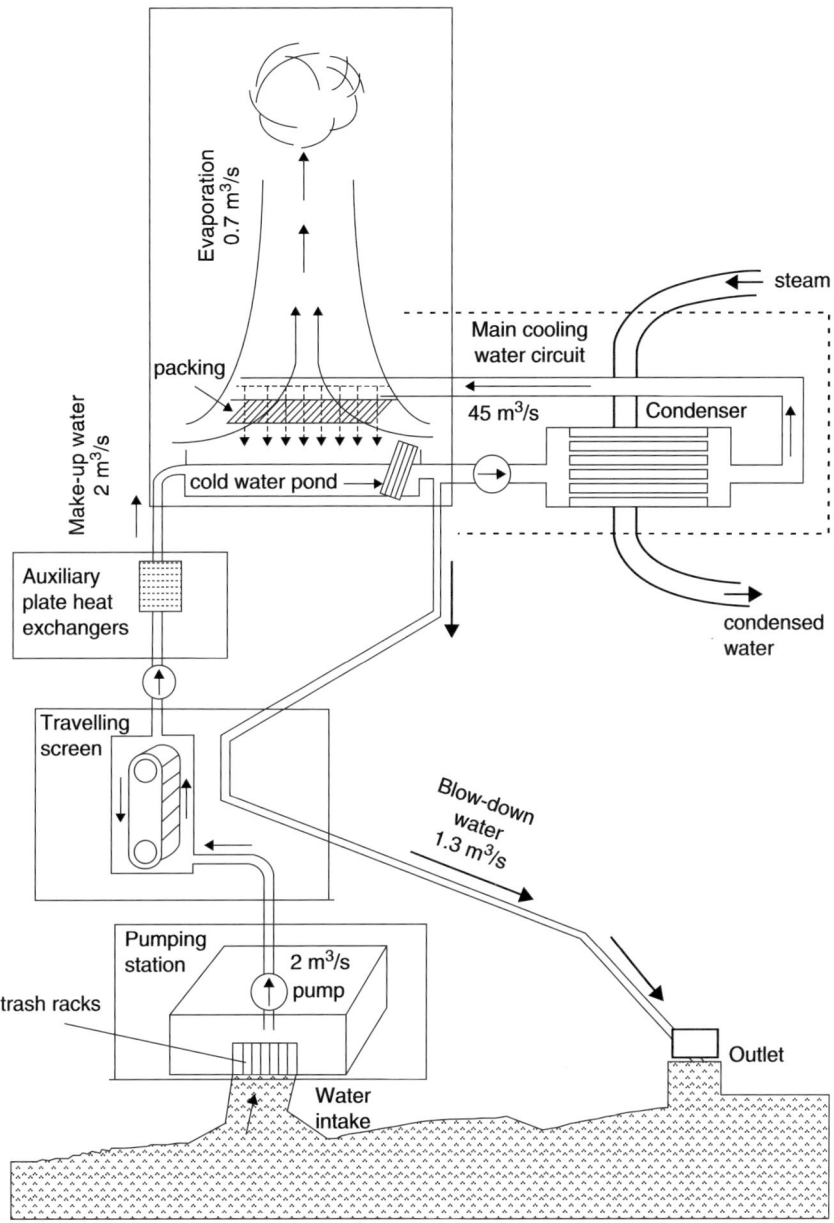

Fig. 5.5. Schematic representation of an open recirculating system. (From Jenner et al., 1998.[1])

the heat exchanger(s) system is cooled down in a cooling tower, where the majority of the heat is discharged to the environment (Figs 5.6 and 5.7). In the cooling tower, the heated water is distributed over the cooling tower fill, and is cooled by contact with air and collected in a reservoir, after which it is pumped back to the reservoir to be reused as a coolant. The air movement is created naturally or by means of fans that push or pull the air through

Fig. 5.6. Cooling water towers at Didcot Power Station, UK. (Photo courtesy of Pisces Conservation Ltd.)

Fig. 5.7. Aerial view of the 2000-MW coal-fired Didcot Power Station showing the six parabolic cooling towers. (Photo courtesy of Pisces Conservation Ltd.)

the tower. Cooling of the water is a result of evaporation of a small part of the cooling water and of sensible heat loss by the direct cooling of water by air, also called convection. The main causes of water loss are evaporation, blowdown (windage, drift, purge (intentional blowdown) and leaks). Intentional blowdown is the draining of water from the circuit necessary to avoid concentration of dissolved solids. To compensate for the blowdown and evaporation, make-up water is added. Generally, the make-up water flow used by an open recirculating system is about 1–10% of the flow of a once-through system with the same cooling capacity. Blowdown generally ranges from 0.15 to 0.80 m³/s per 1000 MWth cooled.

Potential environmental issues
Recirculating system impacts depend on the type of cooling tower and the way it is operated. Impingement and entrainment deaths are reduced to only about 20% or less of that caused by a once-through system. The main impacts are:

- Cooling water additives and their emission through the blowdown to surface water.
- Emissions into air.
- Plume formation, condensation and ice formation.
- Noise.
- Waste due to replacement of cooling tower fill.
- Human health aspects.
- Effects related to the extraction of water, including impingement and entrainment.

Closed (dry) circuit cooling systems

In air-cooled cooling systems (normally termed dry cooling systems) the turbine water is circulated through coils, tubes or conduits, which are cooled by a passing air stream (Fig. 5.8).

Closed circuit dry air-cooled systems consist of finned tube elements, coils or conduits of a condenser, fans with drives and a carrying steel construction or a tower. The process medium itself or a coolant (indirect system) is circulated through the tubes. An air stream is created, naturally or by fans, that flows past the tubes, thus cooling the medium by conduction and convection. If the process medium is a fluid, the cooling system is called an air-cooled fluid cooler. If it is a vapour that is condensed to liquid, the cooling system is called an air-cooled condenser.

Potential environmental issues
Impingement and entrainment deaths are reduced to negligible levels. The main impacts are:

- Noise.
- Impingement of aerial plankton.
- Health issues.

Fig. 5.8. Dry cooling system at the Bethlehem Power Plant, New York. (Photo courtesy of Dr Richard Seaby, Pisces Conservation Ltd.)

Combined wet/dry cooling systems

The open wet/dry cooling tower or hybrid cooling tower is a special design that has been developed as a solution to the problems of cooling water use and plume formation.

It combines both 'wet' and 'dry' cooling tower features: evaporative and non-evaporative cooling. A hybrid cooling tower can be operated either as a pure wet cooling tower or as a combined wet/dry cooling tower, depending on the ambient temperature. The heated cooling water first passes through a dry section of the cooling tower, where part of the heat load is removed by an air current, which is often induced by a fan. After passing the dry section, water is further cooled in the wet section of the tower, which functions in a way similar to an open recirculating tower. The heated air from the dry section is mixed with the vapour from the wet section in the upper part of the tower, thus lowering the relative humidity before the air current leaves the cooling tower, which reduces plume formation above the tower.

Potential environmental issues
The major difference between a hybrid cooling tower and a conventional cooling tower is that the hybrid cooling tower uses 20% less water than a wet cooling tower. There is therefore a proportional reduction in entrainment and impingement.

Definitions of Impingement and Entrainment

Impingement is used here to describe the capture of fish and other organisms on the filter screens of a cooling water intake system (Fig. 5.9). These organisms are washed off the screens and either collected in a trash basket for subsequent disposal or are sluiced along a channel and returned to the environment. Even when a return system is installed, it will not ensure that fish and other organisms survive (Fig. 5.10). Survival depends on the vulnerability of the organism to damage when it comes into contact with a hard surface. Open water fish, such as the herring family, generally have low survival following impingement, because their skins are easily damaged.

Entrainment is a term used here to describe the fate of organisms that are drawn via the cooling water intake structure into the cooling system. The organisms pass through filter screens, travel along the plant's pipework, and are discharged back to the environment with the heated effluent water. Of particular concern is the entrainment of fish eggs and larvae, which may be killed in very large numbers during passage through a power plant's condensers (Fig. 5.11). Recent studies show that mortality rates of entrained organisms can be as high as 97%, depending on the species and life stage entrained. It is often assumed that 100% mortality occurs.

Fig. 5.9. Rotating drum screens at a once-through seawater system. (Photo courtesy of Dr Richard Seaby, Pisces Conservation Ltd.)

Fig. 5.10. Impinged fish, mostly sprats washed from a rotating filter screen into a collecting basket for disposal. Open water fish such as sprat do not survive passage through fish return systems. (Photo courtesy of Pisces Conservation Ltd.)

Entrainment and Impingement with Different Cooling Methods

In this section I consider how entrainment and impingement mortality would be reduced if different cooling methods were used. Three main options are compared: (i) once-through, as used at present in many older plants; (ii) evaporative cooling towers; and (iii) closed-cycle dry cooling. A description of each of these cooling systems was given above.

Fig. 5.11. Planktonic organisms entrained in a power station and collected after passage through the condenser circuit. (Photo courtesy of P.A. Henderson.)

Predicting impingement

Two key features affect the number of organisms killed by impingement: location and volume of water pumped.

The locality and nature of a water body influence the rate of impingement by determining the local abundance of fish and other large aquatic organisms. Fish are not randomly distributed, so the geographical locality and position in the water of the intake can have a large effect on the number of fish and other creatures captured. It is therefore essential that impingement rates are calculated for the appropriate habitat. However, it is not possible to subdivide habitats too finely or we will have insufficient impingement data within each subdivision to produce a reliable description. Separate estimates of the volume extracted–fish impingement relationship have been calculated for: (i) rivers and lakes; (ii) US Great Lakes; and (iii) marine and estuarine habitats.

Within a single water body, the larger the volume pumped, the larger the number of organisms that will be impinged. Living animals, particularly the larger fish and crustaceans that are powerful swimmers, do not behave like passive objects and thus their catch rate can vary non-proportionally with the volume of water pumped. As will be shown below, catch rates increase as a power function of the volume of water extracted. A particularly extreme example of a non-linear increase in impingement with the volume pumped was the study by Wyman (1984)[2] at

Lake Ontario power plants operating with different numbers of cooling water pumps. He found that *Alosa pseudoharengus* and *Osmerus mordax* were apparently attracted to the water currents entering the intake and were caught in greater numbers per unit volume as the volume pumped increased. This response has often been observed, but it is usually explained by increased intake velocities leading to more fish entering a zone where water speed exceeds their sustainable swimming speed, or that the zone where velocity exceeds sustainable swimming speed covers a wider area. In contrast, Wyman found that *Morone americana*, *Morone chrysops*, *Dorosoma cepedianium* and *Perca flavescens* were caught at a constant rate per unit volume irrespective of flow, and *Micropterus dolomieui* were caught in lower numbers per unit volume as flow increased. It was concluded that this latter species avoided faster flowing waters and was thus proportionately more vulnerable to intakes with a reduced pumping rate.

The predicted calculations presented below use US data, which are the best data available and cover a wide range of plants.

Impingement–flow relationship for unprotected intakes

Annual impingement estimates were collected from the literature for ten power plants drawing their cooling water from freshwater rivers (Table 5.1). To be included in the analysis, impingement data had to be collected over at least 1 year to ensure full seasonal coverage. This is because fish impingement is often highly seasonal, varying with breeding season and migration. The sources of these data are listed in Table 5.2.

Using regression analysis, the best fit to a simple function gave the equation:

$$I = 2 \times 10^{-10} V^{4.7592} \qquad (1)$$

where I is the number of fish impinged per year and V the volume extracted in cubic feet per second.

This equation and the impingement data are plotted in Fig. 5.12.

In similar fashion, regression analysis for the best fit to the combined freshwater, estuarine and marine data gave the equation:

$$I = 0.4719 V^{1.8699} \qquad (2)$$

Quantitative prediction of impingement for different cooling water configurations

Using the estimated impingement from equations (1) and (2) allows an estimate for an intake with no fitted fish protection technologies to reduce impingement. To illustrate the effects of various cooling water options, calculations are presented for the Potomac River Generating Station, USA, with average water usage of 350 million gal/day (1.3 million m³/day). This is approximately 541 ft³/s.

The estimated annual impingement using the river relationship (eqn 1) with an average flow of 541 ft³/s is therefore 20,364 fish.

The estimated annual impingement using the all plant relationship (eqn 2) with an average flow of 541 ft³/s is therefore 60,905 fish.

Table 5.1. The number of fish impinged at freshwater and marine power plants in the USA. (Data supplied by Pisces Conservation Ltd.)

		Ref.	Number of fish impinged	Average flow (ft³/s)	% Clupeid/anchovy
Marine/	San Onofre (SONGS)	1696	5,900,000	3,698	
Estuarine	Indian Point	SPDES	1,172,374	2,945	5.00
	Roseton	SPDES	166,395	1,929.6	40.00
	Bowline	SPDES	74,070	1,273.6	6.00
	Danskammer	SPDES	359,781	405.4	34.20
	PH Robinson	287	7,195,785	2,009.6	
	Brunswick, Palico Estuary	287	2,206,000	2,088	
	Calvert Cliffs	286	1,930,000	4,898.4	52.30
	Crystal River, Florida	89	62,028	670	
	Morro Bay	1681	95,800	623.8	70.00
	Maine Yankee	1688	11,134,000	1,002.6	10.00
	Big Bend	316b	212,289	1,977	18.1
	Brayton	316b	51,424	1,546	34.4
	Contra Costa	316b	334,357	518	37.9
	Pilgrim	316b	35,425	691	10.9
	Pittsburgh	316b	430,003	703	20.67
	Salem	316b	8,887,032	4,089	37.4
	Seabrook	316b	10,194	896	7.20
	Salem harbor	1688	12,000	980.3	
	Millstone	1688	21,703	999.9	10
	North Point (Long Island)	1688	18,839	1,044.9	10
	Astoria	1688	215,016	3,288.5	67
	Edge Moor	1688	942,193	1,675.7	
	Surry James River Estuary	1688	4,301,466	3,743	47
	Brunswick	1688	1,303,829	2,896.4	10
	A M Williams	1688	59,936	550.3	62
	Cedar Bayou	1688	150,187,770	1,503.9	98
	Barney M Davis	1688	3,975,182	757.5	85
Freshwater Rivers	Beckjord	316b	46,959	893	
	Cardinal	316b	163,593	1,279	71.30
	Clifty Creek	316b	1,727,393.5	1,829	63.50
	Kammer	316b	12,520	815	90.96
	Kyger	316b	186,223	1,625	65.01
	P Sporn	316b	52,136	1,263	71.70
	Tanners	316b	179,492	1,294	38.80
	WH Sammis	316b	381,173.4857	1,833	79.60
	Quad City Station Mississippi	946	2,428,588	1,810.3	78.99
	Buck Steam Station	508	4,069	580.8	84.00
Lakes	Allen Stream Station Lake Wylie	508	898,913	917	
	Oconee Nuclear	508	1,064,262	4,006.7	
	Marshall Steam Station	508	3,769,300	1,782.1	

Table 5.2. Sources of data used in impingement and entrainment model. (Data supplied by Pisces Conservation Ltd.)

Plant name	Document
San Onofre (SONGS)	Murdoch, W.W., Fay, R.C. *et al.* (1989) Entrapment of juvenile and adult fish at SONGS. Technical report C. Final report of the Marine Review Committee to the California Coastal Commission
Indian Point, Roseton, Bowline	DEIS - SPDESP for Bowline Point, Indian Point 2 & 3 and Roseton Steam Generating Plants
Danskammer	NA (2001) Roseton and Danskammer Point Generating Stations: Impingement Monitoring Program 2000 Annual Progress Report. Normandeau Associates Inc., Poughkeepsie, New York
PH Robinson, Brunswick	Public Service Electric and Gas Company (1976) Draft environmental statement related to the construction of Atlantic Generating Station Units 1 and 2, US Nuclear Regulatory Commission
Calvert Cliffs	Martin Marietta Corporation (1977) Summary of current findings: Calvert Cliffs Nuclear Power Plant Aquatic Monitoring Program, Martin Marietta Corporation
Crystal River, Florida	Grimes, C.B. (1975) Entrapment of fish on intake water screens at a steam electric generating station. *Chesapeake Science* 16(3), 172–177
Morro Bay	CEC (2002) Morro Bay Power Plant Project, California Energy Commission
Big Bend, Brayton, Contra Costa, Pilgrim, Pittsburgh, Salem, Seabrook, Beckjord, Cardinal, Clifty Creek, Kammer, Kyger, P Sporn, Tanners, WH Sammis	EPA 316 b docket for existing plant
Maine Yankee, Salem harbor, Millstone, North Point, Astoria, Edge Moor, Surry, A M Williams, Cedar Bayou, Barney M Davis	Stupka, R.C. and Sharma, R.K. (1977) Survey of fish impingement at power plants in the United States. *Vol. III Estuaries and Coastal Waters.* Argonne, Illinois, Argonne National Laboratory, 310
Quad City Station Mississippi	Latvaitis, B., Bernhard, H.F. *et al.* (1976) Impingement studies at Quad-Cities Station, Mississippi River. Third National Workshop on Entrainment and Impingement
Buck, Allen, Oconee, Marshall	Hanson, C.H., White, J.R. *et al.* (1977) Entrapment and impingement of fishes by power plant cooling water intakes: an overview. *Marine Fisheries Review* 39(10), 7–17
Douglas Point	Kelso, J.R.M. and Leslie, J.K. (1979) Entrainment of larval fish by the Douglas Point Generating Station, Lake Huron, in relation to seasonal succession and distribution. *Journal of the Fisheries Research Board of Canada* 36, 37–41
Unknown Great Lake Power Plants	Kelso, J.R.M. and Milburn, G.S. (1979) Entrainment and impingement of fish by power plants in the Great Lakes which use the once-through cooling process. *Journal of Great Lakes Research* 5(2), 182–194

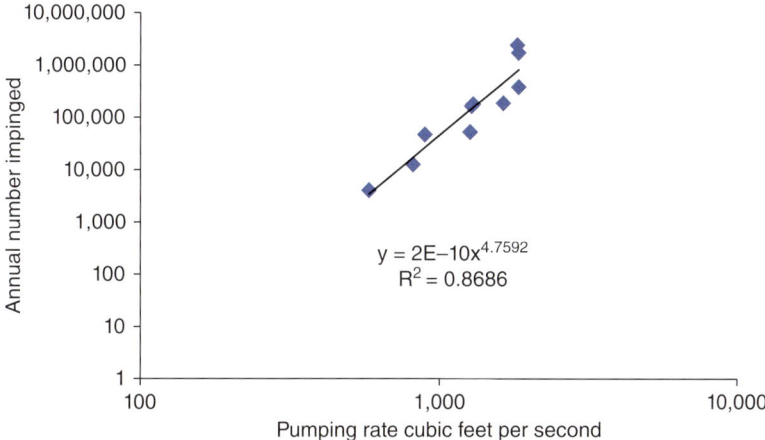

Fig. 5.12. The annual rate of impingement of fish in relation to the volume of water pumped per unit time at freshwater power stations in the USA. (From P.A. Henderson.)

If an evaporative cooling water system was used, the water usage would be a maximum of 10% of that used by once-through cooling. Therefore, the once-through flow of 541 ft³/s would be reduced to a maximum of 54 ft³/s.

Using eqns 1 and 2 above, I can again calculate the number of fish impinged.

The estimated annual impingement with an average flow of 54 ft³/s is therefore from less than 1 to about 800 fish, depending on the flow relationship used.

Essentially, the equation predicts that the flow is so small most fish can avoid impingement. In practice, this is certainly the case, as an intake of this size can be protected using a fine wedgewire screen with a low through-screen velocity that will ensure no fish enter the plant.

The effects of the various options on impingement mortality are summarized in Table 5.3.

Predicting entrainment of young fish

Annual entrainment of fish in US power plant cooling water systems was estimated using the same approach to that developed for impingement above.

As with impingement, the two key aspects that affect the number of young fish killed by entrainment are location and size of the intake. Within a single water body, the larger the volume pumped, the larger the number of organisms that will be entrained. The locality and nature of a water body will influence the abundance of fish eggs and larvae. Some localities are particularly favoured as spawning grounds, while other regions, such as highly turbid waters in muddy estuaries, are not. The available entrainment data are shown in Table 5.4. Note the extraordinary magnitude of these numbers means they are quite difficult to comprehend.

Table 5.3. Summaries of annual impingement rates of fish at the Potomac River Generating Station under different cooling water options. (Data supplied by Pisces Conservation Ltd.)

Intake and cooling method	Total estimated impingement deaths	Notes
Once-through cooling without fish exclusion technology fitted	20,000 to 61,000 per year	The assumed flow is 541 ft³/s or 350 million gal/day. This is the average water usage prior to the 24 August 2005 shutdown.
Once-through cooling with all screens fitted with 'fish buckets' and fish return system	16,000 to 49,000 per year	The survival depends on the species, and ranges from zero for herring and shad to around 40% for white perch. A value of an overall 20% survival has been assumed.
Evaporative cooling	Less than 1 to 800 fish per year	If an evaporative cooling water system were used, the water usage would be a maximum of 10% of that used by once-through cooling. Therefore the once-through flow of 541 ft³/s would be reduced to a maximum of 54 ft³/s.
Evaporative cooling with fine mesh intake screens	Zero	It is assumed at a 3-mm wedgewire screen with a maximum velocity through the screen of less than 0.5 ft/s is installed.
Dry cooling	Zero	There is no or negligible water extraction from the river.

Previous studies have separated source waters for analysis into two components: (i) freshwaters of all types, and (ii) marine and estuarine. The Potomac River, which will be used as a worked example below, holds a major spawning area for both white perch and herring about 9 miles (14 km) above the Potamic Power Plant at Little Falls. Further, given that the locality of the plant is classified as tidal freshwater, this would suggest that the entrainment estimate from ocean and estuarine power plants should be used. However, the resident fish community clearly is freshwater in composition. The approach taken is therefore to calculate entrainment assuming both estuarine and freshwater conditions, and using the order of magnitude estimate of entrainment.

Entrainment–flow relationship for unprotected intakes
Annual entrainment estimates for fish were collected from the literature for 14 power plants drawing their cooling water from freshwaters including the Great Lakes (Table 5.4). To be included in the analysis, entrainment data had to be collected over at least 1 year to ensure full seasonal coverage. The sources of these data are listed in Table 5.2.

Using regression analysis, the best fit to a simple function gave the equation:

$$En = 6 \times 10^7 V^{0.0446} \tag{3}$$

Table 5.4. The number of fish entrained at freshwater and marine power plants in the USA. (Data supplied by Pisces Conservation Ltd.)

		Ref.	Number of fish entrained	Average flow ft³/s	% Clupeid/ anchovy
Marine/ Estuarine	Big Bend	316b	131,076,273,923	1,977	50.45
	Brayton	316b	13,039,270,203	1,546	64.52
	Contra Costa	316b	181,432,719	518	9.80
	JR Whiting	316b	1,182,989,518	288	48.13
	Monroe	316b	1,800,603,000	2,439	78.82
	Pilgrim	316b	3,923,990,671	691	2.24
	Pittsburgh	316b	243,784,289	703	5.77
	Salem	316b	12,947,770,126	4,089	96.30
	Seabrook	316b	838,261,170	896	60.50
	Indian Point	SPDES	702,781,429	2,945	79.00
	Roseton	SPDES	1,205,816,667	1,930	49.50
	Bowline	SPDES	124,082,857	1,274	91.00
	Albany	SPDES	1,090,000,000	552	
	Astoria	SPDES	204,000,000	804	10.00
	Morro Bay	SPDES	508,228,000	624	
Freshwater Rivers	Beckjord	316b	59,727,031	893	3.50
	Cardinal	316b	78,952,478	1,279	0.70
	Clifty Creek	316b	70,076,436	1,829	0.40
	Kammer	316b	182,203,778	815	1.20
	Kyger	316b	589,408,478	1,625	0.80
	P Sporn	316b	113,880,478	1,263	0.90
	Tanners	316b	11,810,166	1,294	0.80
	WH Sammis	316b	17,364,677	1,833	1.20
	Miami Fort	316b	156,247,687	99	12.40
Lakes	Douglas Point	937	17,360,000	301	
	Unknown Great Lake	453	79,706,900	1,375	
	Unknown Great Lake	453	587,774,000	1,041	
	Unknown Great Lake	453	136,110,000	3,512	
	Unknown Great Lake	453	25,896,440	167	

where En is the number of fish entrained per year and V the volume extracted in cubic feet per second. This equation and the entrainment data are plotted in Fig. 5.13. It is clear that the relationship is poor and there is little change in entrainment with flow. Essentially, for freshwater stations, the annual entrainment estimate is of the order of 10^8 individuals (100 million) over a wide range of different intake flow rates. This reflects the importance of locality and proximity to spawning grounds in determining the entrainment numbers.

Annual entrainment estimates were collected from the literature for 15 power plants drawing their cooling water from estuarine or marine waters (Table 5.4). To be included in the analysis, entrainment data had to be collected over at least 1 year to ensure full seasonal coverage.

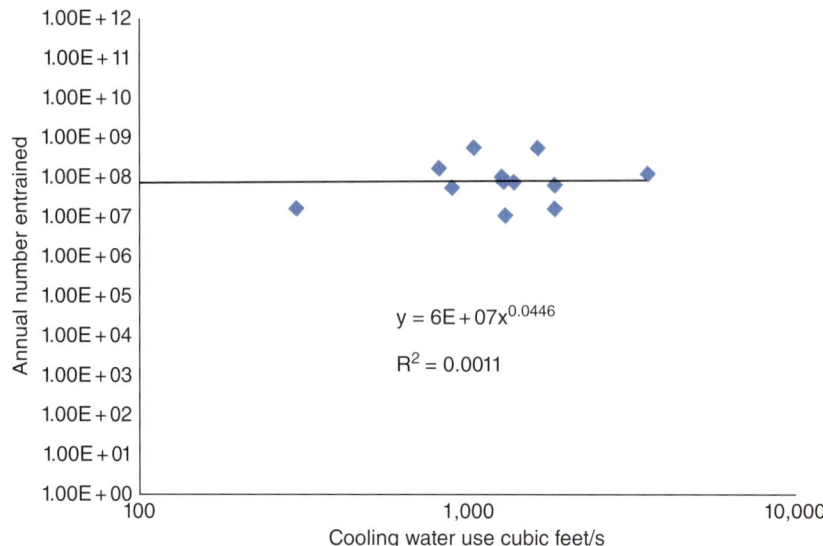

Fig. 5.13. Total annual entrainment in relation to flow for freshwater stations in the USA. (From P.A. Henderson.)

Using regression analysis, the best fit to the combined marine and estuarine data gave the equation:

$$En = 562938V^{1.1144} \tag{4}$$

where En is the number of fish entrained per year, and V the volume extracted in cubic feet per second.

This equation and the entrainment data are plotted in Fig. 5.14. There is a wide variation between sites, with evidence for a positive relationship between flow and entrainment.

Quantitative prediction of entrainment for different cooling water configurations

Using the estimated entrainment from eqns 3 and 4 gives an estimate for an intake with no fitted fish protection technologies to reduce entrainment. This is effectively the present situation.

To illustrate the effects of various cooling water options, I will again use calculations for the Potomac River Generating Station with average water usage of 350 million gal/day (1.3 million m³/day). This is approximately 541 ft³/s.

The estimated annual entrainment using the freshwater relationship with an average flow of 541 ft³/s is therefore 7.9×10^7 fish.

The Potomac River Generating Station is actually placed in the region defined as tidal estuarine; it may therefore be more appropriate to include data from estuarine and ocean sites in the model because of the predominance of anadromous fish.

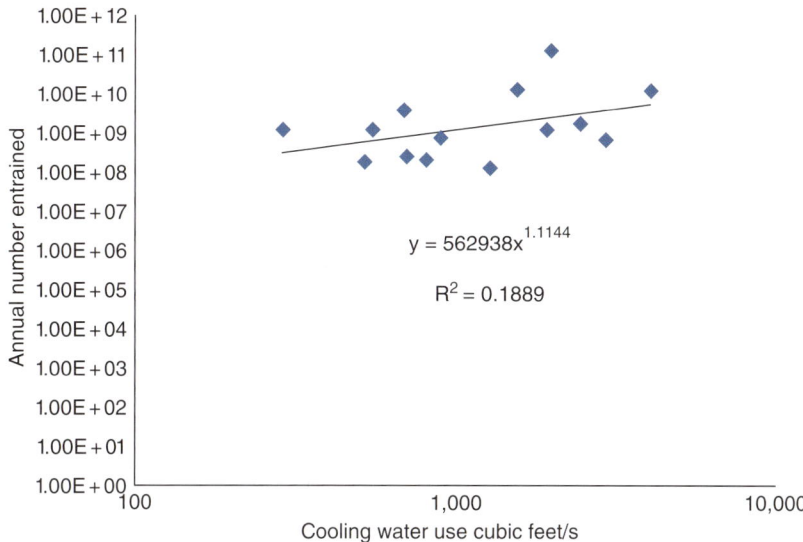

Fig. 5.14. Total annual entrainment in relation to flow for marine and estuarine stations in the USA. (From P.A. Henderson.)

The estimated annual entrainment using the ocean and estuarine relationship with an average flow of 541 ft³/s is therefore 6.25×10^8 fish.

If an evaporative cooling water system were used, the water usage would be a maximum of 10% of that used by once-through cooling. Therefore, the once-through flow of 541 ft³/s would be reduced to a maximum of 54 ft³/s.

Using eqns 3 and 4 above, I can again calculate the number of fish entrained.

The estimated annual entrainment with an average flow of 54 ft³/s is therefore from about 4.8×10^7 to about 7.16×10^7 fish, depending on the flow relationship used.

The only approach to appreciably reduce the entrainment of the eggs and larvae of fish is to cover the intake with a fine mesh screen. Recently, EPRI (2006)[3] reported field trials of the effectiveness of wedgewire screens to reduce entrainment. The level of protection offered depends on the species present, the slot width of the screen and the velocity through the screen. It summarized its findings as follows: 'For all species of larvae combined, the 0.5 mm screen reduced entrainment by 72 and 58 percent, respectively, at slot velocities of 0.15 and 0.30 m/s. The reduction provided by the 1.0 mm screen was 36 (0.15 m/s) and 53 (0.30 m/s) percent.' For the present analysis, I assume a 1-mm slot width working at optimal performance so that a 50% reduction in larval entrainment is achieved.

The effects of the various options on entrainment mortality of fish are summarized in Table 5.5.

Table 5.5. Summaries of annual entrainment rates of fish at the Potomac River Generating Station under different cooling water options. (Data supplied by Pisces Conservation Ltd.)

Intake and cooling method	Total estimated entrainment	Notes
Once-through cooling without fish exclusion technology fitted	In the region of 0.8×10^8 to 6.0×10^8 young fish stages per year	The assumed flow is 541 ft³/s or 350 million gal/day. This is the average water usage prior to the 24 August 2005 shutdown.
Once-through cooling with the intake fitted with a 1-mm wedgewire filter screen system	In the region of 0.4×10^8 to 3.0×10^8 young fish stages per year	For the present analysis I assume a 1-mm slot width working at optimal performance so that a 50% reduction in larval entrainment is achieved. NB: There may be constraints on the ability to install a large wedgewire screening system.
Evaporative cooling	In the region of 4.8×10^7 to about 7.16×10^7 young fish stages per year	If an evaporative cooling water system were used, the water usage would be a maximum of 10% of that used by once-through cooling. Therefore the once-through flow of 541 ft³/s would be reduced to a maximum of 54 ft³/s.
Evaporative cooling with fine mesh intake screens	In the region of 2.4×10^7 to about 3.6×10^7 young fish stages per year	For the present analysis I assume a 1-mm slot width working at optimal performance so that a 50% reduction in larval entrainment is achieved.
Dry cooling	Zero	There is no or negligible water extraction from the river.

In Combination Impacts of Once-Through Cooling

One feature which power plant operating companies avoid is an analysis of the combined effects of all the power plant on the habitat. Once-through cooling has a particularly large impact when all the intakes and outfalls affecting a habitat are combined. I could use as an example power plants situated on the American Great Lakes or along the Hudson River Estuary, New York. The chosen example is for an even greater region, the waters surrounding Great Britain, which include some of the finest fishing waters on the planet. This is an interesting example because it includes power plants in a number of countries which generally do not share information on impingement and entrainment and have a history of not even acknowledging the problem.

In the late 1990s there were 45 large, direct-cooled power stations bordering Britain and its adjacent continental coast. For the 119 species of fish killed during cooling water extraction, it was estimated that 3 to 5×10^8 individuals >3-cm long are killed per annum. For eggs, larvae and fish

<3-cm long, the mortality rate is about 10^{14} individuals per annum. For some species, these losses may significantly reduce abundance or undermine conservation efforts. Fish are particularly vulnerable because power stations are often sited in estuarine nursery areas or on migratory paths. The 17 power stations sited in the southern North Sea are estimated to kill sole and herring equivalent to about 50% of British commercial landings for the region. In Northern Europe, there is a need for international cooperation to determine the magnitude of fish kills and their effect on populations.

From the 1950s there was a steady development of coastal direct-cooled power stations which use large volumes of cooling water (about 50 m^3/s for a 1250-MW nuclear power station). The fate of animals sucked in with the water depends on their size. Fish >30 mm in length are caught by screens of 2.5–10 mm mesh. These impinged fish are usually killed, even if a return system has been installed. Those which pass through the screens travel via the station condenser circuits to be discharged to sea. Few survive passage because they suffer mechanical, temperature, biocides and pressure damage.

Data on passage and impingement mortalities were either obtained from the literature or estimated by extrapolation from adjacent power stations for which surveys had been undertaken. Most estimates refer to plant operation in the 1980s. Similar, if not larger, losses are probably still occurring. Table 5.6 lists all the studies known to have been undertaken in Northern Europe and gives the estimated total annual kill through impingement for each station under the assumption that all the installed capacity is in constant use. Almost all inshore fish species are vulnerable to capture by cooling water intakes. The stations listed in Table 5.6 have caught 119 of the 122 inshore species known from the region (Henderson, 1989).[4] For the 33 stations for which estimates could be obtained, the annual kill is estimated as 5.7×10^8 individuals >3-cm long. These stations comprise 83% of the total installed pumping capacity. Plant availability and demand are constantly changing, but probably 50–75% of the total pumping capacity is used per annum, giving a total annual mortality on the filter screens of between 3 and 5×10^8 individuals. The commonest species caught are sprat *Sprattus sprattus*, whiting *Merlangius merlangus*, sand goby *Pomatoschistus* spp. and flounder *Platichthys flesus*. Table 5.6 also includes estimates of the number killed passing through the station. Because of lack of data for many stations, reliable estimates of total mortality cannot be given, but indications are that total mortality for eggs and fish <3-cm long are of the order of 10^{10} and 10^{14} individuals per annum, respectively.

Dramatic as these numbers are, they have little meaning unless compared against either population size or mortality rates from other causes. While isolated populations in the vicinity of intakes may be destroyed, it is difficult to determine if mobile, widely distributed fish are also likely to be suffering significant losses. Only for commercial fish are there estimates of population size or mortality rate to compare power station kills against. To gain some appreciation of the significance of the numbers, a

Table 5.6. Estimated number of fish killed on the filter screens of marine and estuarine large power stations situated on the European coasts of the southern North Sea and North East Atlantic. Data collected from the late 1970s onwards. This table gives the scale of the problem at plants with once-through cooling. (For sources, see Henderson, 1989, 2017.[8, 9])

Name	Type, cooling water volume m³/s, screen mesh mm	Total number of fish killed on the filter screens per annum Fish >3 cm	Total number of species recorded on the filter screens	Method of estimation of impingement data or source	Total number of fish eggs passing through the condenser circuits per annum	Total number of young fish passing through the condenser circuits per annum Fish <3 cm	Method of estimation of passage data or source
Inverkip	Conventional, 11.5, 10						
Peterhead	Conventional, 28, 10						
Hunterston B	Nuclear, 30, 10						
Methil	Conventional, 3.4, 10	Insignificant					
Longannet	Conventional, 90, 10						
Kincardine	Conventional, 11.5, 10						
Cockenzie	Conventional, 38, 10						
Torness	Nuclear, 45, 8	3.0×10^4	39	1-year survey 1991–1992 (Anon., 1993)			
Blyth A & B	Conventional, 55, 10	8.6×10^6	49	3-year survey 1979–1981, qualitative data only (Davis and Dunn, 1982) Quantitative estimate extrapolated from Hartlepool			
Hartlepool	Nuclear, 26, 9.5	4.0×10^6	35	3-year survey (Peaty, 1993)			
Sizewell A	Nuclear, 34.4, 9.5	3.7×10^6	73	1-year survey 1981–1982 (Turnpenny et al., 1983)	2×10^{10}	4.9×10^7	Unpublished survey by Fawley Aquatic Research Laboratories 1992–1993

Sizewell B	Nuclear, 48, 9.5	5.2×10^6		Extrapolation Sizewell A – plus survey and comparative data for 1994	3.3×10^{10}	8.0×10^7	Extrapolation Sizewell A
Bradwell	Nuclear, 26, n/a	3.20×10^6		Extrapolation Sizewell A	1.7×10^{10}	4.2×10^7	Extrapolation Sizewell A
Tilbury C	Conventional, 50, 10	4.7×10^6		Extrapolation from W. Thurrock		3.7×10^{13}	Extrapolation from Kingsnorth
W. Thurrock	Conventional, 58.4, 10	4.9×10^6	68	10-year survey		3.7×10^{13}	Extrapolation from Kingsnorth
Littlebrook D	Conventional, 50, 10	4.7×10^6		Extrapolation from W. Thurrock		3.7×10^{13}	Extrapolation from Kingsnorth
Kingsnorth	Conventional, 64, 9.8	9.9×10^5	59	2-year survey 1977–1979 (van den Broek, 1980)		4.7×10^{13}	2-year survey (Dempsey, 1983)
Isle of Grain (half operating)	Conventional, 57.6, 9.8	9.9×10^5		Extrapolation from Kingsnorth		4.7×10^{13}	Extrapolation from Kingsnorth
Dungeness A	Nuclear, 27, 9.5	7.4×10^5		Extrapolation from Dungeness B	$>3.5 \times 10^8$	$>4.3 \times 10^8$	Extrapolation from Graveline
Dungeness B	Nuclear, 40, 9.5	1.1×10^6	79	1-year survey unpublished	$>5.6 \times 10^8$	$>7 \times 10^8$	Extrapolation from Graveline
Fawley	Conventional, 60, 10	6.0×10^5	80	1-year survey 1973–1974 (Holmes, 1975)	1.23×10^6	2.4×10^7	1-year survey 1986–1987 (Dempsey, 1988)
Hinkley Point A	Nuclear, 40, 10	1.3×10^6		Extrapolation from Hinkley B			

Continued

Table 5.6. Continued.

Name	Type, cooling water volume m³/s, screen mesh mm	Total number of fish killed on the filter screens per annum Fish >3 cm	Total number of species recorded on the filter screens	Method of estimation of impingement data or source	Total number of fish eggs passing through the condenser circuits per annum	Total number of young fish passing through the condenser circuits per annum Fish <3 cm	Method of estimation of passage data and source
Hinkley Point B	Nuclear, 30, 10	9.9×10^5	73	16-year survey unpublished			
Oldbury-Upon-Severn	Nuclear, 26.5, 10	2.5×10^5	75	5-year survey 1971–1976 (Claridge et al., 1986)			
Berkeley	Nuclear, 26.5, 10	2.5×10^5	71	Qualitative data (Claridge et al., 1986), quantitative estimation by extrapolation from Oldbury-on-Severn			
Uskmouth	Conventional, 30.3, 5	2.9×10^5	35	Qualitative data (Claridge et al., 1986), quantitative estimation by extrapolation from Oldbury-on-Severn			
Aberthaw B	Conventional, 67, 10	2.2×10^6		Extrapolation from Hinkley B			
Pembroke	Conventional, 50, 10	1.6×10^6		Qualitative data (Claridge et al., 1986), quantitative extrapolation from Hinkley B			

Station	Type						
Wylfa	Nuclear, 68.3, 8	4×10^4	59	1-year survey unpublished	9.66×10^7	1.08×10^8	2-year survey 1986–1987 (Dempsey and Rogers, 1989)
Heysham 1	Nuclear, 33.3, 10	7.7×10^5	51	1-year survey unpublished Extrapolation Heysham 1			
Heysham 2	Nuclear, 50, 10	1.6×10^6					
Coolkeeragh	Conventional, 11.5, 10	17×10^5	28	1-year survey, 1989–1990 (Moorehead and Service, 1992)			
Ballylumford	Conventional, 29.4, 8	1.06×10^5	41	1-year survey, 1989–1990 (Moorehead and Service, 1992)			
Belfast West	Conventional, 9.1, 10	1.5×10^4	27	1-year survey, 1989–1990 (Moorehead and Service, 1992)			
Kilroot	Conventional, 16.6, 5	1.1×10^5	37	1-year survey, 1989–1990 (Moorehead and Service, 1992)			
Moneypoint Graveline	Conventional, 40, 10 Nuclear, 240, 2.5	2.14×10^8	49	Estimated from 2-year survey 1981–1982 (Blanpied-Wohrer, 1984)	$>3.22 \times 10^9$	$>4 \times 10^9$	Estimated from 2-year survey 1981–1982 (Blanpied-Wohrer, 1984)

Continued

Table 5.6. Continued.

Name	Type, cooling water volume m³/s, screen mesh mm	Total number of fish killed on the filter screens per annum Fish >3 cm	Total number of species recorded on the filter screens	Method of estimation of impingement data or source	Total number of fish eggs passing through the condenser circuits per annum	Total number of young fish passing through the condenser circuits per annum Fish <3 cm	Method of estimation of passage data and source
Dunkirk	Conventional, 21, 3	6.2×10^5	32	Électricité de France unpublished data	$>2.8 \times 10^8$	$>3.5 \times 10^8$	Extrapolation from Graveline
Paluel	Nuclear, 160, 3	2.04×10^9 (1984) 2.7×10^8 (1985)	46	2-year survey unpublished study by Électricité de France	$>2.14 \times 10^9$	$>2.66 \times 10^9$	Extrapolation from Graveline
Flamanville	Nuclear, 80, 3						
Maasvlakte	Conventional, 11.5, 10	1.0×10^7		1+ years survey			
Eems	Conventional, 55, ?						
Borssele	Conventional, 34.5, ?						
Doel	Nuclear, 50, 3	2.5×10^7		1-year survey (Maes et al., 1996)			

series of different approaches are taken for six widespread species, each representing a different lifestyle. Sole (*Solea solea*) is an abundant, commercially fished, benthic flatfish with planktonic eggs and young that settle close inshore. Herring (*Clupea harengus*) is pelagic, has benthic eggs and has been commercially over-exploited. The eel (*Anguilla Anguilla*) is an abundant catadromous species breeding in the Sargasso Sea. Bass (*Dicentrarchus labrax*) ranges from Northern Scotland to the Mediterranean and the juvenile fish use estuaries. Twaite shad (*Alosa fallax*) is an anadromous member of the herring family that was common and widely distributed in Northern European estuaries but is now considered endangered.

Sole are abundant in the southern North Sea (ICES area IVc), where there is a commercial fishery. Seventeen large power stations were operating in this region in the 1980s and if all were working at full capacity the calculated Equivalent Adult Value (EAV) of the sole they would kill by number and weight would be approximately 1.73×10^6 and 3.37×10^5 kg, respectively. These estimates are only for the egg and post-metamorphosis stages because there are no quantitative data for the capture of larval sole on these power stations. The EA weight is 45.9% and 10.5% of average UK and International sole landings for area IVc for the years 1989–91 and 1986–88, respectively (MAFF fisheries statistics). If account is taken of intermittent power station operation, the number killed for the years 1985–1992 would probably be reduced to between 50% and 75% of the maximum. However, it is known that larval sole are caught at some stations and it is certain that not all small sole killed were observed because some pass through the screens.

Previous studies of UK east coast herring estimated an EAV of 435 tonnes per annum, which was 50% of the average annual UK commercial landings for ICES area IVc between 1989 and 1991 (Turnpenny and Henderson, 1992).[5] The EAV for total power station mortality would be higher because this estimate did not include passage mortality. However, the largest captures of North Sea herring probably occur at power stations in the English Channel, including Dungeness, Paluel and Graveline. Single filter screens at Paluel and Graveline were estimated to catch 1.55×10^7 and 1.16×10^7 O-group herring per annum, respectively (Unpublished Électricité de France report). Running at full capacity, both stations catch in the order of 10^8 O-group herring per year. Scottish stations are also situated in important herring nursery areas resulting in kills of 10^6–10^7 post-larval herring per annum.

The eel (*Anguilla anguilla*) supports a small commercial fishery for which no landing statistics are available. This common and widely distributed fish has declined dramatically in abundance in many European rivers over the last 40 years. Unlike most fish, it is caught both by stations sited on rivers and the coast. Elvers and glass eels are caught, but passage mortality is unknown. The number of large individuals caught is known. For the 18 power stations included in Table 5.6 for which reliable data on eel impingement were available, the number killed between 1980 and 1990 was about 2.4×10^5 per annum. By extrapolation, the total annual catch of maturing eels by coastal European stations in Northern Europe is of the order of 10^6 individuals. They are also killed by freshwater power stations. A study of fish impingement at six Dutch freshwater stations gave an estimated total annual mortality of 1.17×10^5 (Hadderingh *et al.* 1983).[6]

Following concern about declining bass catches in British waters during the 1980s, inshore estuarine nursery areas were defined where fishing for bass is restricted. No restrictions were placed on the operation of power stations in nursery areas. For Kingsnorth Power Station on the Medway Estuary it is estimated that during the winter of 1987/88 2–5% of the local O-group population was killed on the filter screens (Pickett and Pawson, 1994).[7] Grain, W. Thurrock, Littlebrook and Tilbury power stations were also situated within the Thames basin. Kingsnorth and W. Thurrock killed an estimated 7.7×10^3 and 43.5×10^3 bass per annum respectively, indicating that mortality at W. Thurrock and the adjacent Littlebrook may have had a more important impact on the bass population. Another important nursery is the Severn Estuary. During the 1980s, Berkeley and Oldbury power stations operated in the centre of distribution of the young bass. Berkeley is now closed. The estimated annual catch at Oldbury was 1.49×10^5 individuals; Berkeley was probably catching similar numbers. Other power stations operating in the Severn Estuary/ Bristol Channel and capturing juvenile bass were Hinkley Point A and B (2.7×10^3 per annum), Uskmouth, Aberthaw and Pembroke. It can be argued that the MAFF policy to protect nursery areas was flawed and perhaps of negligible benefit because it imposed no regulation on the power companies, the single largest killer of juvenile bass. Similar arguments could be made for continental countries. Paluel Power Station in France, for example, catches an estimated 2.16×10^5 per annum at full capacity. For the 18 power stations included in Table 5.6 for which reliable quantitative data on bass impingement were available, the number killed between 1980 and 1990 was about 4.7×10^5 per annum.

European power stations also impact on endangered species such as twaite shad (*Alosa fallax*), which is listed under schedule 5 of the Wildlife Conservation Act and restricted in Britain to only four breeding populations. The largest British population is in the Bristol Channel–River Severn, where Hinkley Point A, B and Oldbury power stations killed an average of 4.17×10^4 individuals per annum during the 1980s. In this region, smaller kills also occurred at Berkeley, Uskmouth, Aberthaw and Pembroke.

We still have insufficient knowledge to determine if cooling water intakes are having a measurable effect on inshore fish abundance. The full extent of the problem is larger than presented because no estimates are given for other cooling water intakes such as those on petrochemical plants. Fish deaths can be reduced by measures such as fish deterrent and return systems, modified intake design and seasonal switching of production. This problem is not unique to Europe and high fish kills have been a concern in countries such as the USA.

The USA has taken a lead in reducing entrainment and impingement losses by banning once-through cooling unless it can be demonstrated to produce levels of mortality equivalent to closed-cycle cooling. Unfortunately, this is not the case in Europe. In Britain, for example, the Environment Agency has actually argued that once-through cooling is the best technology available. This decision has allowed the new 3200-MW Hinkley C Nuclear Power Station to be designed with a once-through cooling system pumping between 116 and 134 m^3/s. It is claimed that they can reduce fish mortality by the design of the intake and the use of a fish return system. Both are unproven and highly unlikely to reduce the mortality significantly. There is an almost unsurmountable technical problem building a fish return system with a tidal range of more than 12 m because the fish cannot be returned easily during low water periods.

A Review of Technologies for Protecting Fish from Impingement and Entrainment

To measure the effectiveness of impingement and entrainment reduction technologies, a standard intake configuration is required to compare them against. This is assumed to be the level of impingement and entrainment at a cooling water intake equipped solely with conventional travelling screens designed to prevent debris and adult and juvenile fish from entering the condenser tubes. Such intakes typically have trash racks usually consisting of fixed bars to prevent large debris from entering the system, and drum or band screens, usually with 3/8-inch mesh in the USA, that prevent smaller debris from clogging the condenser tubes. The screens are rotated and washed intermittently at a typical pressure of 80 to 120 pounds per square inch.

Modified travelling screens with fish return – Ristroph screens

The simplest modification to normal conditions is to run the screens continuously and use a fish return sluice to return impinged fish to the water. Simply returning an impinged fish to the river does not ensure that it survives. This was shown by the EPRI (2007)[10] study at the Potomac River Generating Station, the main findings of which are summarized below.

EPRI (2007) reported on the impingement survival of about 3000 fish impinged on screen 1A of the Potomac River Generating Station in 2005 and 2006. The survival rates after 28 h were highly species-specific and varied over time. Delicate-bodied open water fish such as shad and river herring did not survive impingement. However, bluegill and eel had survivals of 100% or just below. The most abundant species impinged over the study period was white perch, *Morone americanus*, which had 28-h survival of 56% and 30% in 2005 and 2006, respectively. This considerable difference between years was related to the capture of fish during storm events when large amounts of debris were also impinged. High debris loadings were associated with reduced survival. As will be shown below, these findings are consistent with those from other studies.

Studies reported by Muessig *et al.* (1988)[11] undertaken at Danskammer, Bowline Point and Roseton power plants on the Hudson showed that 84–108-h survival was highest for Atlantic tomcod, striped bass and white perch, at about 50–90%, and lowest for bay anchovy, alewife and blue-back herring at 0–25%. Other species such as pipefish and centrarchids (sunfishes) had survival rates between these extremes. These same general features indicating large differences between fish families were also recorded in studies of the fish return system at Sizewell nuclear power station on the English east coast, where 24-h survivals were from about 100% for flatfish such as flounder and plaice, 40–90% for cod, bass and whiting, to 0% for Atlantic herring and sprat. These studies and others indicate that fish return systems fitted to unmodified screens will not aid the survival of herring family (clupeid) fish to any appreciable extent.

Ristroph screens have water-filled lifting buckets that collect the impinged organisms and transport them to a fish return system. The buckets are designed such that they will hold approximately 2 inches of water when above the water. The buckets hold the fish in water until the screen rises to a point where the fish are spilled onto a bypass, trough or other protected area (Mussalli *et al.*, 1978).[12] Fish baskets are another modification and may be used in conjunction with fish buckets. Fish baskets are separate framed screen panels that are attached to vertical travelling screens. An essential feature of modified travelling screens is continuous operation during periods when fish are being impinged. Conventional travelling screens typically operate on an intermittent basis. Impinged fish are usually returned to the source water body by sluiceway or pipeline.

It is inevitable that fish will try to avoid capture in the lifting buckets and will have a high risk of coming into contact with hard surfaces. This will result in exhaustion and damage to their skin, fins and eyes, which can be sufficiently severe to lead to the mortality of the more fragile species. The degree of mortality will depend on the species and the particular life stage, plus numerous physical factors including water temperature, oxygen concentration and salinity. It is generally the case that open water (pelagic) species, which are poorly adapted for contact with hard surfaces, are particularly vulnerable to surface damage to their bodies. Thus, it is inevitable that the reduction in impingement mortality that these screens can produce when compared against a conventional travelling screen standard is variable and species-specific.

To determine overall performance requires information on species-specific mortality rates and the species composition of the impinged community. Ideally, we would require data on the temporal variation in both the fish community composition and mortality rates, but as these are unavailable a degree of uncertainty must be accepted. This uncertainty is noted in the recent EPA review (Efficacy of Cooling Water Intake Structure Technologies, Chapter 3),[13] although this gives a notably precise and high estimate of the reduction in fish mortality: 'In summary, performance data for modified screens and fish returns are somewhat variable due to site conditions and variations in unit design and operation. However, the above results generally show that at least 70–80 percent reductions in impingement can be achieved over conventional travelling screens.'

A key aspect to consider when analysing fish survival data from Ristroph screens is the time after impingement and handling when survival was measured. Some early studies quoted high survival after 10–15 minutes in a holding tank. This is clearly of little interest, because most acutely-injured fish will take considerably longer to die. The minimum time at which survival rates are likely to give a true indication of the eventual survival of the impinged fish will be after about 8 h. Fletcher (1990)[14] gives estimates for the survival of common species at Indian Point in the Hudson estuary after this time period (Table 5.7).

The survivals presented in this table are probably typical. The Kintigh Generating Station, New Jersey, has recorded survivals of generally greater than 80% for rainbow smelt, rock bass, spottail shiner, white bass, white perch and yellow perch. Gizzard shad survivals have been 54–65% and alewife survivals have been 15–44%.

The above and similar results have been highly influential in guiding the EPA to the conclusion that Ristroph screens could achieve at least 70–80% reductions in mortality. However, it should be noted that there are a number of factors that will likely reduce further the survival of fish.

First, survival over an extended period will be lower than the 8-h survival. It has been found that stressed and damaged fish can take a number of days to die. Experiences in angling and fish farming demonstrate that quite minor damage may lead to bacterial and fungal infections resulting in eventual death. There is also the problem with all fish return systems that exhausted, disorientated and damaged individuals can be picked off by predators on their return to the main water body. It is normal to observe large predatory fish and piscivorous birds waiting and feeding at water discharges.

The progressive decline in survival with time following impingement is demonstrated in data collected at Roseton Generating Station in the Hudson estuary (Table 5.8). Apart from spottail shiner, all other species showed a marked decline in the rate of survival between 2.5 and 96 h after impingement. This clearly indicates the need to use survival estimates

Table 5.7. Eight-hour survival rates for Indian Point. (From Fletcher, 1990.[14])

Fish species	Survival %
Bay anchovy	77
American shad	65
Blueback herring	74
Striped bass	91
White perch	86
Atlantic tomcod	83
Alewife	38

Table 5.8. Data from 1994 impingement mortality studies at Roseton (dual flow screens). (From NAI, 1995.[15])

Species	Number	Survival rate over time					
		0 h	2.5 h	8 h	24 h	48 h	96 h
American shad	575	0.689	0.252	0.123	0.080	0.071	0.068
Alewife	1839	0.662	0.229	0.151	0.096	0.073	0.060
Bay anchovy	1093	0.282	0.169	0.110	0.032	0.014	0.004
Blueback herring	8973	0.753	0.335	0.204	0.110	0.090	0.071
Striped bass	899	0.889	0.740	0.578	0.494	0.405	0.345
Spottail shiner	331	0.958	0.931	0.915	0.897	0.873	0.831
White perch	899	0.950	0.909	0.828	0.727	0.648	0.583

over periods of at least 96 h if the post-impingement survival is to be estimated correctly.

Second, temperature and salinity can have a great effect on survival. Injured fish are far more likely to die at low temperatures and at low salinities. Salinity is probably important because damage to the skin results in a loss of osmotic control. The effect of both these variables is demonstrated in the work of Muessig *et al.* (1988).[11] While these studies were carried out using conventional rather than Ristroph screens, they still give insights into the effects of salinity and temperature on injured individuals.

The above considerations indicate that clupeid fish such as the alewife and American shad are more vulnerable to damage than fish belonging to other families. Further, short-term survival rates at intermediate water temperatures and salinities are unlikely to fully reflect the eventual mortality rate for species that are easily injured.

The above considerations suggest that the post-impingement survival rates presented in the PSEG Power New York Inc.'s *Bethlehem Energy Center SPDES Modification, Alternative Cooling Systems Study for Ristroph Screens* (Table 5.9) give a fair appraisal of survival for American east coast estuarine and marine fish. Further, these values can provide survival estimates for fish families that can be used to make general estimates of post-impingement survival with Ristroph screens for species for which no data exist.

Calculated total survival rates for Ristroph screens

Using the post-impingement survival rates given in Table 5.9 in conjunction with data on annual impingement, it is possible to calculate the total reduction in impingement mortality that can be achieved by Ristroph screens. For illustrative purposes the calculations for the Albany Steam Generating Station on the Hudson estuary are considered in detail. These are used because of the quality of the data.

ALBANY STEAM GENERATING STATION. The estimated annual number of animals killed while using conventional and Ristroph screens is presented for each

Table 5.9. The post-impingement survival of fish with conventional and Ristroph screens. (From PSEG Power New York Inc.'s Bethlehem Energy Center SPDES Modification, Alternative Cooling Systems Study.)

Family	Species	Per cent survival	
		Conventional	Ristroph type
Acipenseridae	Atlantic sturgeon	60	80
	Shortnose sturgeon	60	80
Anguillidae	American eel	70	95
Bothidae	Summer flounder	70	95
Catostomidae	White sucker	50	70
Centrarchidae	Black crappie	30	40
	Bluegill	80	80
	Largemouth bass	75	90
	Longear sunfish	70	80
	Pumpkinseed	75	80
	Redbreast sunfish	70	80
	Rock bass	70	80
	Smallmouth bass	75	90
	White crappie	30	40
Clupeidae	Alewife	0	10
	American shad	0	10
	Blueback herring	0	10
	Gizzard shad	5	10
	AW/BBH	0	10
Cyprinidae	Bluntnose minnow	50	90
	Carp	50	80
	Common shiner	50	90
	Creek chub	50	90
	Emerald shiner	50	90
	Fallfish	50	90
	Golden shiner	45	90
	Goldfish	50	80
	Rosyface shiner	50	90
	Silvery minnow	50	90
	Spotfin shiner	50	90
	Spottail shiner	50	90
	Unidentified shiner	50	90
Cyprinodontidae	Banded killifish	85	90
	Mummichog	85	90
Engraulidae	Bay anchovy	0	80
Esocidae	Chain pickerel	70	90
	Northern pike	70	90
	Redfin pickerel	70	90
Gadidae	Atlantic tomcod	10	70
Gasterosteidae	Fourspine stickleback	70	90
	Threespine stickleback	70	90
Ictaluridae	Brown bullhead	65	90
	Channel catfish	70	90
	Tadpole madtom	70	90

Continued

Table 5.9. Continued.

| Family | Species | Per cent survival | |
		Conventional	Ristroph type
	White catfish	75	90
	Yellow bullhead	70	90
Osmeridae	Rainbow smelt	0	85
Percichthyidae	Striped bass	25	70
	White bass	25	70
	White perch	25	70
Percicidae	Logperch	65	80
	Tessellated darter	90	100
	Walleye	65	80
	Yellow perch	65	80
Percopsidae	Trout-perch	15	20
Petromyzontidae	Lamprey spp.	70	95
Salmonidae	Brown trout	60	80
Sciaenidae	Freshwater drum	20	25
Soleidae	Hogchoker	90	95
Umbridae	Central mudminnow	60	80

species in Table 5.10. The impingement using conventional screens is estimated by sampling the operating screens. The estimate with Ristroph screens fitted is made by multiplying the annual total with conventional screens by the predicted survival rate for each species as estimated from data in Table 5.9. The total number of fish killed by impingement with and without Ristroph screens is predicted to be 311,636 and 198,074, respectively. Thus, the total percentage reduction in the number of fish killed at this site would be (311,636–198,074) × (100/311,636) = 36.4%. While many species were predicted to have very high survivals, the average was much lower because of the dominance at this locality of clupeid fish with low survival rates. The tendency for clupeids to bring down overall survival rates is an almost global feature.

Cylindrical wedgewire screens

Wedgewire screens have a proven ability to reduce both impingement and entrainment mortality at low volume intakes (1–50 MGD). Their effectiveness is related to the: (i) slot width; (ii) through-slot velocity; (iii) existence and strength of ambient cross flow to carry organisms away from the screen; (iv) amount of biofouling; and (v) amount of ambient debris. As the EPA note, they are an unproven technology for protecting once-through intakes that typically pump volumes in excess of 100 MGD.

Wedgewire screens with slot widths of 5–10 mm have been used to effectively eliminate impingement at freshwater cooling water intakes.

Table 5.10. Calculations of the number of individuals predicted to be killed by impingement after the installation of Ristroph screens, using data for the Albany Steam Plant, New York. (Data supplied by Pisces Conservation Ltd.)

Species	Total impingement	Proportion surviving (%)	Total killed
Atlantic sturgeon	2	80	0.4
Shortnose sturgeon	40	80	8
American eel	6,904	95	345.2
Summer flounder	1	95	0.05
White sucker	159	70	47.7
Black crappie	220	40	132
Bluegill	3,213	80	642.6
Largemouth bass	52	90	5.2
Longear sunfish	2	80	0.4
Pumpkinseed	1,160	80	232
Redbreast sunfish	75	80	15
Rock bass	280	80	56
Smallmouth bass	52	90	5.2
White crappie	25,840	40	15,504
Alewife	18,458	10	16,612.2
American shad	11,779	10	10,601.1
Blueback herring	141,033	10	126,929.7
Gizzard shad	4,535	10	4,081.5
AW/BBH	67	10	60.3
Bluntnose minnow	29	90	2.9
Carp	76	80	15.2
Common shiner	4	90	0.4
Creek chub	3	90	0.3
Emerald shiner	90	90	9
Fallfish	1	90	0.1
Golden shiner	585	90	58.5
Goldfish	324	80	64.8
Rosyface shiner	6	90	0.6
Silvery minnow	603	90	60.3
Spotfin shiner	1,233	90	123.3
Spottail shiner	22,988	90	2,298.8
UID shiner	16	90	1.6
Banded killifish	92	90	9.2
Mummichog	3	90	0.3
Bay anchovy	321	80	64.2
Chain pickerel	3	90	0.3
Northern pike	1	90	0.1
Redfin pickerel	33	90	3.3
Atlantic tomcod	377	70	113.1
Fourspine stickleback	2	90	0.2
Threespine stickleback	2	90	0.2
Brown bullhead	359	90	35.9
Channel catfish	13	90	1.3
Tadpole madtom	2	90	0.2
White catfish	2,229	90	222.9

Continued

Table 5.10. Continued.

Species	Total impingement	Proportion surviving (%)	Total killed
Yellow bullhead	25	90	2.5
Rainbow smelt	151	85	22.65
Striped bass	4,618	70	1,385.4
White bass	12	70	3.6
White perch	59,741	70	17,922.3
Logperch	3	80	0.6
Tessellated darter	1,957	100	0
Walleye	3	80	0.6
Yellow perch	1,411	80	282.2
Trout-perch	75	20	60
Lamprey spp.	1	95	0.05
Brown trout	7	80	1.4
Freshwater drum	1	25	0.75
Hogchoker	309	95	15.45
Central mudminnow	55	80	11
Total	311,636		198,074

To reduce entrainment of fish eggs and larvae appreciably, the screen slot widths need to be in the range 0.5 to 3.0 mm. Weisberg *et al.* (1987)[16] found that a 3 mm slit width excluded about 50% of bay anchovy and naked goby larvae in the 5–6 mm long size class. A 1-mm slot width gave almost complete exclusion of bay anchovy greater than 8 mm in length and naked goby greater than 7 mm long. To give good protection to the very small larvae, a slot width of 0.5 mm is required.

A 0.5 mm slot width will only be highly effective for larval exclusion when used with a suitable intake velocity. At a velocity of 7.5 cm/s, this width will exclude larvae less than 6 mm in length. However, at a velocity of 15.0 cm/s (0.5 ft/s), about 60% of larvae less than 7.0 mm in length were entrained.

The reduction of egg entrainment is related to the size of the egg. However, eggs are not rigid and eggs greater than 0.5 mm in diameter will pass through a 0.5 mm slot. Data on the entrainment of marine fish eggs through a 0.5 mm slot width screen with a velocity of 7.5 cm/s are presented in Table 5.11.

A species of particular importance in many US estuaries is the striped bass. This species has a relatively large egg (2.4–3.9 mm diameter) and thus egg entrainment would almost certainly be eliminated by slot widths in the range 0.5–1.0 mm. However, the striped bass yolk sac larvae range in length from 2–7 mm, which would suggest that some young larvae would be entrained with even a 0.5 mm slot width, and very limited protection would be offered by a width >1.0 mm.

An estimate of the level of entrainment exclusion that could be achieved with a velocity of 7.5 cm/s through the screen would be in the order of 90% for a 1 mm screen width in flowing rivers and 80–85% for estuarine sites with a 0.5 mm screen width.

Table 5.11. Entrainment of marine fish eggs through a 0.5 mm slot width screen with a velocity of 7.5 cm/s. (From Sunset Energy Facility proposal for Brooklyn, New York).

Species	Egg diameter mm	% Exclusion
Tautog	0.7–1.14	80
Bay anchovy	0.65–1.24	84–75
Windowpane flounder	1.0–1.4	96–93
Atlantic menhaden	1.0–2.0	100
Weakfish	0.9–1.1	100

Fine mesh travelling screens

The incidence of entrainment can be greatly reduced by the use of 0.5–1 mm mesh travelling screens. However, this does not mean that the mortality of young fish is proportionately reduced, as the eggs and early stages are now liable to impingement damage. Taft *et al.* (1976)[17] report laboratory studies of the effects of impingement on fine mesh screens for the larval stages of striped bass, winter flounder, alewife, yellow perch, walleye, channel catfish and bluegill. Survival was highly variable and dependent on water velocity through the screen and the duration of impingement. The highly species-specific nature of survival of impinged larvae was also noted by McLaren and Tuttle (1999).[18]

Fletcher (1990)[14] also noted that the mortality on fine mesh screens is related to the amount of debris retained by the screen. This would suggest that fine mesh screens would not be effective in all waters. Fletcher (1992)[19] reports a study of the effectiveness of fine mesh screens to reduce losses of early life stages of striped bass. The results showed that survival was influenced by mesh size, water velocity and exposure time. It was concluded that impingement resulted in high mortality for young larvae and many larvae that initially survived impingement subsequently died. The results suggested that striped bass up to 8.4-mm long are too delicate to survive impingement.

Given the high maintenance of fine screens together with the known high impingement mortalities of many species, these devices cannot be considered a useful protective measure.

Barrier nets

Under appropriate conditions, barrier nets can be effective devices to reduce fish impingement. To be effective there must be limited debris in the water, a low incidence of biofouling, relatively low water velocities and sheltered conditions with low wave action, low current velocities, etc. In estuarine conditions, their effectiveness is questionable. The following is described for the barrier net deployment at Bowline Point Generating Station on the Hudson Estuary (EPA 316b for existing plant studies):

> The Bowline Point Station (New York) has an approximately 150-foot barrier net in a v-shape around the intake structure. Testing during 1976 through 1985 showed that the net effectively reduces white perch and striped bass impingement by 91 percent. Based on tests of a 'fine' mesh net (3.0 mm) in 1993 and 1994, researchers found that it could be used to generally prevent entrainment. Unfortunately, species' abundances were too low to determine the specific biological effectiveness.

This account gives the impression that the 3 mm net was useful for reducing entrainment. In fact, as Lawler, Matusky & Skelly Engineers (1996) report,[20] in 1993 the net clogged with fine suspended silt and sank. In 1994, even when the net was sprayed to remove clogging it fouled with the alga *Ectocarpus*, causing two of the support piles to snap and the evaluation to end. They concluded that 3 mm barrier nets can only be considered an experimental device.

In summary, under appropriate conditions barrier nets are effective at blocking fish passage; they work well providing velocities are low and debris loading is light. Barrier nets used in Lake Michigan were found to be over 80% effective at protecting target species (yellow perch, rainbow smelt, alewife and chub) (Taft, 2000).[21] Fine barrier nets capable of reducing entrainment have not been successful at estuarine sites. They are probably most appropriate for lacustrine intakes.

Microfiltration

The only microfiltration system that has actually been deployed is the Gunderboom at Lovett Generating Station, New York. Overtopping, tunnelling and rips were observed during testing. For example, in the Lovett evaluation report for 1999 it is stated that 'the divers documented a substantial gap along the bottom of the boom. The gap extended along the bottom of the boom for approximately 3 m and ranged in depth from 0.5 to 0.6 m' (LMS, 2000).[22]

Velocity caps

Velocity caps only work on offshore intakes that take their water from, or near, the seabed. To claim that velocity caps have been successful in minimizing impingement is an exaggeration. They have been found to reduce impingement by 50–80% when compared with an unprotected intake. However, it should be noted that this reduction is usually only observed for pelagic species. Other fish and crustaceans may still be caught in large numbers.

Porous dykes

While large porous dykes could be used to filter water to eliminate impingement and substantially reduce entrainment losses in most localities, they are impractical and therefore cannot be considered as a viable technology. The major problem with porous dykes comes from clogging by debris and silt, and from fouling by colonization of animal and plant life. Back flushing, which is often used by other systems for debris removal, is not feasible at a dyke installation.

Louver screens

Louver screen systems work by guiding fish away from an intake into a channel or part of the water body from which water is not extracted. They can therefore only be applied in localities where there is directional flow and suitable by-pass channels can be built. A particular benefit is that they work equally well in light, dark or turbid conditions (Turnpenny, 1988).[23] Louver systems work well in some applications but require large-scale engineering. They do not provide protection for fish early in their life cycles.

Louver systems have been studied with migratory species at hydroelectric facilities in rivers; they have been little studied at steam generating plants and require further large-scale evaluation before their effectiveness can be proven. In rivers, a 70% reduction in impingement may be possible. They can be assumed to give no reduction in entrainment losses.

Behavioural barriers

The principal ideas that have been tried are sound, bubble and light deterrent systems. In almost all cases they have been found to be ineffective in reducing impingement and are assumed ineffective for entrainment. The one possible exception is the use of sound to deter alewife. In tests at the Pickering station in Ontario, poppers were found to be effective in reducing alewife impingement and entrainment by 73% in 1985 and 76% in 1986. Testing of sound systems at the James A. Fitzpatrick station in New York showed similar results: an 85% reduction in alewife impingement and entrainment. At the Arthur Kill station pilot and full-scale high frequency sound tests showed results for alewife comparable to Fitzpatrick and Pickering. Impingement of gizzard shad was also three times less than without the system. No deterrence was observed for American shad or bay anchovy using the full-scale system.

Variable-speed pumps

In a power station cooling water system, pumps are used to draw water in through the intake and circulate it through the plant's cooling system. A single-speed pump generates the maximum designed flow whenever it is operating. It is either on or off. Pumps can, however, be designed to operate at variable speeds so that when set on a lower speed they withdraw less water from the source water body. Using a variable-speed pump within a power plant cooling system would enable the facility operator to reduce flow whenever conditions such as low inlet temperatures and/or low demand dictate that a reduced flow will provide adequate cooling for the amount of electricity generated. In other words, variable-speed pumps give a plant operator the ability to tailor cooling water withdrawals to the minimum amount actually needed for cooling. Thus, the use of variable speed pumps could allow a reduction in flow for a once-through cooling system at times when higher flows are not needed.

Variable-speed pumps do not provide a constant impingement-reduction benefit. They provide a benefit only when the plant is actually able to reduce flows and still generate electricity and meet thermal discharge limitations. Because of the variability in inlet water temperatures and electricity demand, it is difficult to reliably predict the effectiveness of this option for reducing flow (and impingement) on an annualized basis.

Summary of the Effectiveness of the Different Entrainment and Impingement Reduction Devices

Table 5.12 gives a summary of the possible minimum, maximum and mean per cent reductions in fish entrainment and impingement that the possible technologies are likely to produce.

The Most Protective Devices for Different Cooling Systems and Water Types

In order to compare once-through and evaporative cooling against dry cooling, we need to consider which technologies are the best that can be fitted to reduce impingement and entrainment. The particular options will depend on the habitat. Table 5.13 shows the potential reductions in entrainment and impingement possible if the best protective technologies available for each habitat are fitted.

The Effects of Thermal Discharges

Once-through cooling water systems discharge considerable quantities of warm water to the environment. I will use Dungeness A and B Nuclear Power Station, UK, as a typical example of the type of discharge involved. The cooling water is discharged from two closely spaced outlets set in the seabed at 15 m below mean sea level, about 100 m to the seaward of the low water mark (Figs 5.15 and 5.16). The A station had four condenser cooling water pumps, each with a nominal rating of 6.687 m^3/s, giving a maximum discharge rate of approximately 26.8 m^3/s. The water is discharged at approximately 10°C above the inshore ambient temperature. As can be seen in Fig. 5.16, the discharge plumes of

Table 5.12. Likely reduction in mortality rates at cooling water intakes fitted with various entrainment and impingement reduction devices. (Data supplied by Pisces Conservation Ltd.)

Technology	% Reduction in impingement			% Reduction in entrainment		
	Min.	Mean	Max.	Min.	Mean	Max.
Ristroph screens and fish return system	15	30	70	0	0	0
Wedgewire screen with 1-mm slot width and water velocity of 7.5 cm/s	100 Ineffective in ocean waters	100 Ineffective in ocean waters	100 Ineffective in ocean waters	90 (River) 80 (Lake) 50 (Estuary) Ineffective in ocean waters	90 (River) 80 (Lake) 50 (Estuary) Ineffective in ocean waters	90 (River) 80 (Lake) 50 (Estuary) Ineffective in ocean waters
Fine mesh travelling screens	0	0	0	0	10	30
Barrier nets 1-cm mesh	60 Ineffective in ocean waters	70 Ineffective in ocean waters	85 Ineffective in ocean waters	0	0	0
Microfiltration (Gunderboom)	50 Ineffective in ocean waters	90 Ineffective in ocean waters	100 Ineffective in ocean waters	20 20 Estuarine Ineffective in ocean waters	40 30 Estuarine Ineffective in ocean waters	95 40 Estuarine Ineffective in ocean waters
Louver screens	10	40	70	0	0	0

Table 5.13. The best technologies to protect fish against impingement and entrainment in different aquatic habitats. (Data supplied by Pisces Conservation Ltd.)

Water type	Entrainment devices	Potential reduction	Impingement devices	Potential reduction
River once-through	None		Louver screens	70%
River evaporative	Fine wedgewire screen	90%	Fine wedgewire screen	100%
Upper estuarine – once-through	Gunderboom	40% – when deployed	Barrier net/ Gunderboom	80%
Upper estuarine – evaporative	Fine wedgewire screen	90%	Barrier net/ Gunderboom	80%

Dungeness A and B combine to form a single plume. This primary plume protrudes about 500 m offshore into the coastal current. Once entrained by the tidal stream, the discharge moves parallel to the 20 m depth contour and is rapidly dispersed.

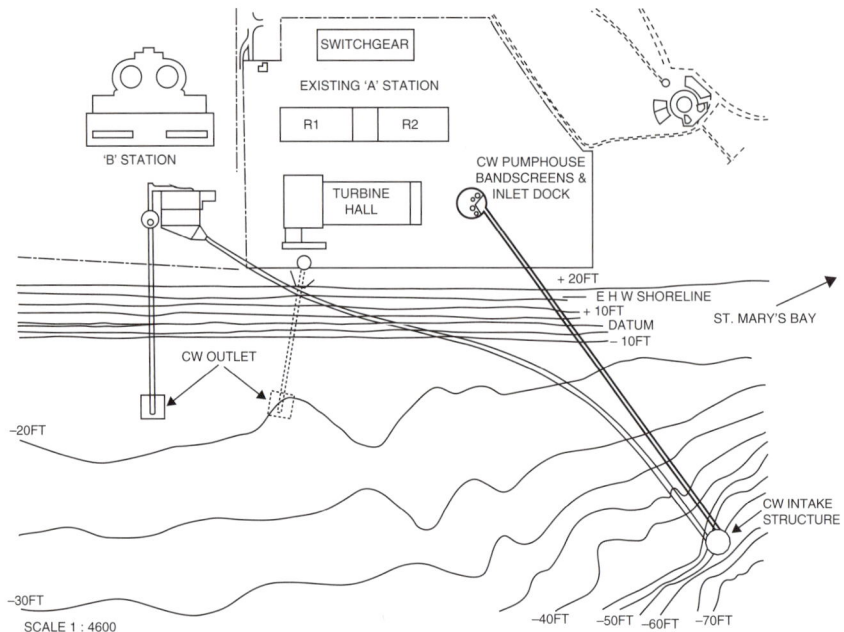

Fig. 5.15. Plan of the Dungeness A and B Power Stations showing coastline and positions of outfalls and intake. (From Pisces Conservation Ltd, from CEGB archive.)

Fig. 5.16. Aerial photograph showing the Dungeness A and B primary plumes at about low water. (Photo courtesy of Pisces Conservation Ltd, from CEGB archive.)

As the density of water declines with increasing temperature, the heated water discharge initially floats on the surface so that the seabed in the vicinity of the point of discharge is not impacted by warmed water (Fig. 5.17).

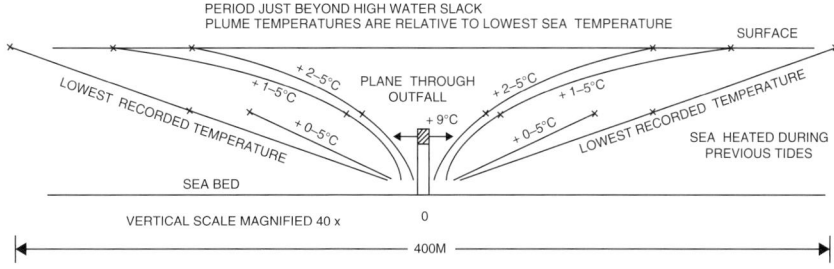

Fig. 5.17. Cross-section of a typical offshore power station discharge plume showing the general change in temperature with depth. Note that in the vicinity of the discharge point the warm water does not impinge the seabed. (This diagram was based on data collected at Sizewell A, Suffolk by Parker (1977).[24])

The effects of thermal discharges on marine life

Planktonic life

PHOTOSYNTHETIC ORGANISMS. In most cases where measurements have been made in receiving waters, the effects of the discharge on phytoplankton have been restricted to the near-field plume, usually within a very short distance of the outfall (Langford, 1990).[25] Generally, it has been difficult to separate chlorination, used to control biofouling, from thermal impacts. For example, intermittent chlorination caused reductions of 80–90% in photosynthesis measured some 50 m from the vertical discharge 'boil' at the San Onofre nuclear power station on the California coast (Eppley *et al.*, 1976).[26] The thermal plume close to the outfall typically remains discrete and, unless the outfall is fitted with diffusers, there is typically little mixing with the receiving water. Thus, low levels of photosynthesis in heated water close to an outfall are related to damage during passage through the condenser circuit.

Few studies have been undertaken on the effects of the plume after appreciable mixing with the receiving water. A notable exception is the study by Smith *et al.* (1974)[27] at the estuarine Indian River Power Station, USA. A regression model was used to estimate rates of photosynthesis at points 2.5 km upstream and downstream of the discharge canal mouth, mostly within the areas bounded by the 0.2°C isotherm. When natural water temperatures were around 5°C, and in the absence of chlorination, the increase in the rate of photosynthesis ranged from 3% to 260% over a ΔT range of 0.2–8°C. Elevated rates were predicted over areas of 2 km upstream and 2–3 km downstream. In summer, when natural temperatures were around 25°C, the predicted changes in photosynthesis ranged from −7% in the discharge canal to +3% at the 2°C isotherm.

This study would suggest that within the 2°C isotherm an increase of 3% in summer photosynthetic activity could be anticipated. While the winter percentage increase may be larger, it will be comparatively insignificant as photosynthetic activity during the winter is a very small proportion of summer rates.

Planktonic invertebrates

Close to the discharge point, where temperatures are most elevated, impacts on planktonic crustaceans have been noted. For example, in a study at Dunkerque Power Station, Brylinski (1981)[28] found a close correlation between temperature and the rate of development of the copepod *Temora longicornis*.

While changes in growth and development can be anticipated, within the area with temperatures elevated by 2°C or more there is little evidence that thermal discharges increase mortality. More recent studies by Bamber and Seaby (2004)[29] found that a ΔT of between 8.3 and 10°C had no significant effect on the common calanoid copepod *Acartia tonsa*. These authors do note, however, that chlorination can cause high mortalities to planktonic animals. Similarly, Bamber and Seaby (2004) showed that larval *Crangon crangon* exposed to a ΔT of about 8°C or more had a significantly higher 48-h mortality rate than that of the control.

Fish

As early as the 1930s, Bull (1936)[30] demonstrated, from a range of marine species covering a number of taxa (not salmonids) and ecotypes, that fish could detect and respond to a temperature front of 0.03 to 0.07°C. Fish will therefore attempt to avoid stressful temperatures by actively seeking water at the preferred temperature, but this becomes increasingly a matter of chance once coordination begins to break down. If an uncoordinated fish is moved to cooler water it may recover, but the chances of recovery decrease with duration of exposure.

At less than stressful levels, increasing temperatures allow increased rates of metabolism and (notably with regard to migratory activity) increased swimming speeds but decreased endurance (Turnpenny and Bamber, 1983; Beach, 1984).[31, 32] The relationship between muscle activity (twitch rate, t), temperature (T, °C) and fish size (length, L), derived from six species, was expressed mathematically by Zhou (1982)[33] as:

$$T = 0.17L^{0.4288} + 0.0028\log_e T - 0.0425L^{0.4288} \times \log_e T - 0.0077$$

The temperature at which locomotor activity becomes disorganized, and thus the fish loses its ability to escape from adverse conditions, has been termed the Critical Thermal Maximum (CTM).

Once temperatures exceed 40°C, heat death ensues: enzymes are inactivated, proteins denature or coagulate and fats melt. The last comprehensive review of this subject, from the molecular to whole organism level, was that of Rose (1967).[34] The response of fish to temperature is complex. Fish have natural thermal niches (preferenda), and in the temperate zone freshwater species are either:

- cold water species, such as salmon, trout, tomcod and smelt;
- cool water species; or
- warm water species, such as carp.

This categorization tends to fall along taxonomic lines in that related species and genera have similar thermal niches (Hokanson, 1977).[35] Cherry *et al.* (2011)[36] found that the stenothermal salmonids had the narrowest temperature tolerance ranges of all the teleosts they studied. Superimposed on this thermal selectivity are temporal variations in preferenda that can be correlated with the age or developmental stage of the fish, its physiological condition or with different environmental variables. Young fish generally have higher thermal preferences and greater tolerances than older fish. Feeding activity, reproductive or migratory behaviour and stress (anoxia, turbidity, salinity changes and chemical pollutants) might substantially alter normal thermal responses.

Some species are better than others at adapting their physiology or behaviour: in general, estuarine species are fairly resilient, since they are subject to regular environmental fluctuation.

For any fish there are temperatures that it prefers, temperatures to which it can acclimate, temperatures that it would seek to avoid but at which it can survive for various periods of time, and temperatures that are lethal. Moreover, the ability of individuals to survive is not the same as the ability of the species to continue to prosper; increased temperatures may advance or delay breeding seasons, encourage breeding in the wrong place or inhibit fish migration.

TEMPERATURE AND DISSOLVED GASES. Indirect effects of temperature on fish include reduced solubility of gases, particularly of oxygen, an effect which can be exacerbated by the elevated temperature simultaneously increasing the rate of oxygen removal by pollutants such as sewage. The sort of temperature elevations that are encountered outside the immediate vicinity of a power station discharge are between $1°$ and $3°C$, which would decrease the solubility of oxygen by only about 0.5 ppm. If the water was 100% saturated with oxygen, this reduction in solubility would lead to outgassing. Gas-bubble disease, caused by nitrogen release (outgassing) within fish with raised temperatures has been recorded in *Oncorhynchus* young, these succumbing in 10–30s in air-saturated water with a ΔT of $1.7°C$ (Snyder, 1969).[37] A similar effect was found by Coughlan (1970)[38] in *Salmo salar* kelts when maintained in supersaturated, heated seawater ($c.13°C$) for 2 months.

THERMAL TOLERANCE OF FISH SPECIES. The effects of temperature on the biology and ecological requirements of fish have been extensively studied and reviewed. Temperature can affect survival, growth and metabolism, activity, swimming performance and behaviour, reproductive timing and rates of gonad development, egg development, hatching success and morphology. Temperature also influences the survival of fish stressed by other factors such as toxins, disease or parasites. Several reviewers have focused on thermal biology, specifically: lethal and/or preference temperatures (e.g. Coutant, 1975; Cherry *et al.*, 2011).[39, 40] Others have widened their reviews to include data on growth, preference and lethal temperatures (e.g. Jobling, 1981).[41] Much of the literature is summarized by Langford (1990).[42]

In open water discharges, fish deaths caused by the temperature of the discharge are unknown; this is almost certainly because of their mobility. There are extensive observations that show that discharges change the distribution of fish. The majority of these indicate that fish are attracted into thermal discharges, although this may not be because of the tempera- ture but rather because of the opportunities to scavenge dead animals re- leased with the discharge.

Benthic life

Algae
If bottom substrates close to a thermal discharge are impacted by the plume, larger algal forms such as *Ascophyllum* spp. and *Fucus* spp. tend to be replaced by *Enteromorpha* and *Cladophora*. At the Maine Yankee Power Station, *Ascophyllum nodosum* and *Fucus vesiculosus* were found to decline in regions where the effluent elevated the temperature between 5 and 7°C above ambient. It can be concluded that, where temperatures exceed 5°C above ambient, large algal species will be reduced or elim- inated and at temperature of 10°C or greater above ambient only simple temperature-tolerant forms will be present.

Invertebrates
An important early study on the effects of a power station discharge on a sandy beach community was undertaken by Barnett (1971)[43] near Hunterston Generating Station, Scotland. He concluded that on the beach adjacent to the outfall, where the temperature was 2–4°C above ambient, the *Tellina tenuis* growth rate increased but population density change could not be linked to the effluent. The growth and seasonality of *Urothoe brevicornis* was also changed by the effluent.

At Kingsnorth Power Station in the Thames Estuary, Bamber and Spencer (1984)[44] found that tidal temperature fluctuations of 10°C caused by the effluent plume stressed the benthic community and resulted in the loss of about 50% of the expected species. Almost all the species pre- sent in the plume were littoral species that were adapted to withstand high summer temperatures when exposed to the sun on beaches (Bamber, 1990).[45] It seems likely that changes in macroinvertebrate growth and sea- sonality will occur when temperatures exceed 2°C above ambient, and large changes in community structure, including a significant loss in diversity, are likely to occur when temperatures exceed 8°C above ambient.

Biofouling Control and Chlorination

The protection of the intake structure and the pipework of a plant from fouling by mussels and other organisms often requires the use of anti-fouling agents.

Bivalve animals, especially mussels, can and do settle and grow in cooling systems; their larvae and juvenile stages pass through intake filter screens. Within the system the animals can cause blockages, while detached mussel shells can cause erosion-corrosion in condenser tubes, thereby threatening plant integrity. Historically mussels had to be cleared by hand from culverts on a regular basis. Many coastal power stations control fouling by chlorination. In freshwaters, a variety of approaches are used including heat treatment. Chlorination products are frequently released into the receiving waters at low levels with the discharge water. Chlorination is important as it will reduce entrainment survival and will kill a high proportion of the bacteria and other micro-organisms present in the intake water.

The change in chlorine concentration through time in a discharge plume

Empirical studies have shown residual chlorine concentration decay in seawater to be a two-stage process. There is a very rapid initial loss of residual chlorine, termed the 'instantaneous chlorine demand', followed by a slower, approximately exponential, decay. To encompass both the chlorine demand and the decay, Davis and Coughlan (1983)[46] developed an empirical mathematical model for the loss of chlorine in a power station discharge plume, of the form:

$$C_t = \frac{(C_{in} - C_{id})e^{-kt}}{(D + k'D)^t}$$

where
C_t is the residual chlorine concentration at time t after discharge (mg/l);
D is the dilution rate;
C_{in} is the chlorine concentration at the point of injection (mg/l);
C_{id} is the instantaneous chlorine demand (mg/l);
k is the chlorine decay constant (per minute);
k' is the chlorine demand constant linked to dilution; and
t is time.

The above equation can be applied if we know the rate of dilution with time, the chlorine decay constant and the instantaneous chlorine demand of the water. The decay constant and instantaneous demand are usually estimated from empirical studies.

The instantaneous chlorine demand

The initial fast loss of residual chlorine from water is termed the instantaneous demand. Davis and Coughlan (1983) measured the instantaneous demand following the introduction of 2.5 mg/l NaOCl. As shown in Fig. 5.18, they found that the instantaneous demand as determined 1 minute after the addition of NaOCl ranged between 0.1 and 1 mg/l for temperatures between 0 and 33°C.

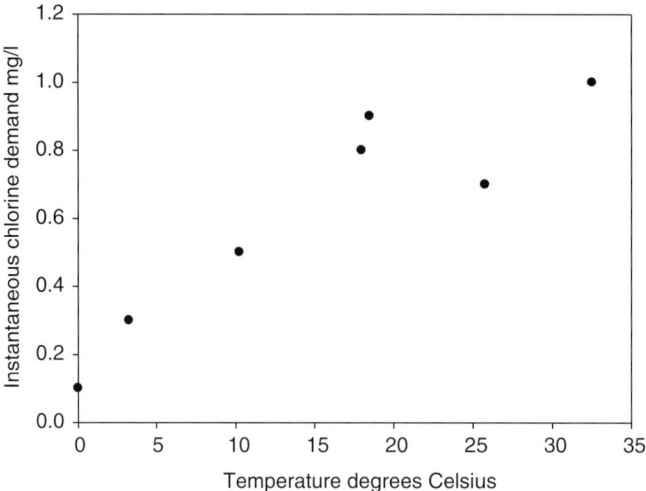

Fig. 5.18. The instantaneous chlorine demand of seawater between 0 and 35°C. The initial chlorine concentration was 2.5 mg/l and the demand was measured 1 minute after addition of the NaOCl. (Based on data in Davis and Coughlan, 1983.[46])

Using non-linear regression, an exponential model of the form

$$C_{id} = a(1 - e^{-bT})$$

gave a good fit to the data ($R^2 = 0.891$), with $a = 0.972$, $b = 0.0888$ and T, the temperature. Therefore, at a temperature of 20°C and a chlorine concentration of 0.25 mg/l, the instantaneous chlorine demand is estimated to be 0.807 mg/l or all of the initial concentration.

Cooling Tower Blowdown

Evaporative cooling towers concentrate the chemicals initially present in the cooling water. There is therefore a need to continually replace a proportion of the cooling water to stop the dissolved solids content of the circulating cooling water from continuously increasing. The discharged water, termed cooling tower blowdown water, can have undesirable properties, such as increased dissolved solids content and an increase in any pollutants present in the make-up water as it is concentrated. It is also normal for cooling tower water to be chemically treated with additives. As a typical example, the cooling tower system at William States Lee III Nuclear Generating Station adds sodium hypochlorite, sodium bromide, sulphuric acid and sodium polyacrylate. In all cases but one, the chemical concentrations in the effluent of the cooling tower system would be non-detectable as they would degrade within the system prior to discharge. Only the silt dispersant, sodium polyacrylate, was identified as being potentially present in the liquid effluent on release.

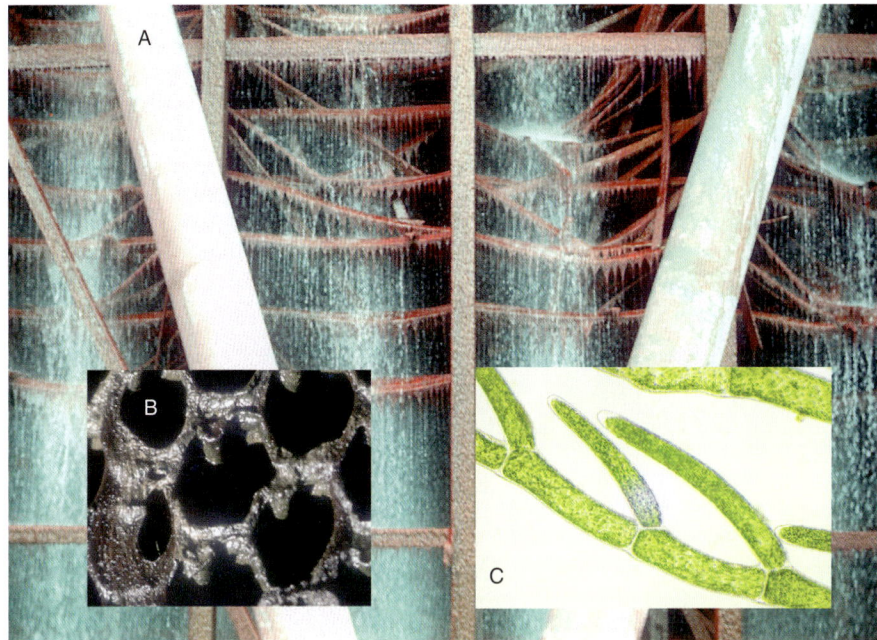

Fig. 5.19. Biofouling of the cooling tower at High Marnham, UK with the filamentous alga *Cladophera* resulting in the collapse of the packing as the weight of the alga increased and silt was retained. (A) View at the base of the cooling tower showing the collapsed packing. (B) Close-up of the packing material showing algal fouling. (C) Micrograph of *Cladophera* filaments. (From Pisces Conservation Ltd, from CEGB archive.)

Evaporative cooling towers are also subject to fouling and can harbour dangerous pathogens requiring the use of biological control methods. Treatments may include antifouling chemicals such as chlorine or the use of heat, resulting in a release of antifouling chemicals or heat to the environment. For example, Fig. 5.19 shows the destructive impact of *Cladophera*, a filamentous alga, at High Marnham Power Station, UK.

Notes

[1] Jenner, H.A., Whitehouse, J.W., Taylor, C.J. and Khalanski, M. (1998) *Cooling Water Management in European Power Stations: Biology and Control of Fouling (Vol. 1).* Electricité de France, Paris.
[2] Wyman, R.L. and Dischel, R.S. (1984) Factors influencing impingement of fish by Lake Ontario power plants. *Journal of Great Lakes Research* 10(4), 348–357.
[3] EPRI (2006) Field Evaluation of Wedgewire Screens for Protecting Early Life Stages of Fish at Cooling Water Intake Structures: Chesapeake Bay Studies, 2006, 1012542. EPRI, Palo Alto, California, USA.
[4] Henderson, P.A. (1989) On the structure of the inshore fish community of England and Wales. *Journal of the Biological Association of the United Kingdom* 69, 145–163.
[5] Turnpenny, A.W.H. and Henderson, P.A. (1992) Sizewell: reappraisal of fish impingement at the 'A' station, and implications of multiple plant operation: Fawley, Research Lab. Report 144.

[6] Hadderingh, R.H., Van Aerssen, G.H.F.M., Groeneveld, L., Jenner, H.A. and Van der Stoep, J.W. (1983) Fish impingement at power stations situated along the Rivers Rhine and Meuse in the Netherlands. *Hydrobiological Bulletin* 17, 129–141.

[7] Pickett, G.D. and Pawson, M.G. (1994) *Sea Bass: Biology and Conservation.* Chapman and Hall, London.

[8] Henderson, P.A. (1989) On the structure of the inshore fish community of England and Wales. *Journal of the Marine Biological Association of the United Kingdom* 69, 145–163.

[9] Henderson, P.A. (2017) Long-term temporal and spatial changes in the richness and relative abundance of the inshore fish community of the British North Sea Coast. *Journal of Sea Research* 127, 212–226.

[10] EPRI (Electric Power Research Institute) (2007) *The Latent Impingement Mortality Assessment of the Geiger Multi-Disc Screening System at the Potomac River Generating Station.* Report 1013065. EPRI, Palo Alto, California.

[11] Muessig, P.H., Hutchison, J.B., King, L.R., Ligotino, R.J. and Daley, M. (1988) Survival of fishes after impingement on travelling screens at Hudson River power plants. *American Fisheries Society Monograph* 4, 170–181.

[12] Mussalli, Y.G, Hofman, P. and Taft, E.P. (1978) Influence of fish protection considerations on the design of cooling water intakes. *Joint Symposium on Design and Operation of Fluid Machinery, Colorado, USA.*

[13] EPA (2014) Technical Development Document for the Final Section 316(b) Existing Facilities Rule. Available at: https://www.epa.gov/sites/production/files/2015-04/documents/cooling-water_phase-4_tdd_2014.pdf (accessed December 2017).

[14] Fletcher, R.I. (1990) Flow dynamics and fish recovery experiments: water intake systems. *Transactions of the American Fisheries Society* 119, 393–415.

[15] Normandeau Associates Inc. (NAI) (1995) *Roseton Generating Station 1994 Evaluation of Post impingement Survival and Impingement Abundance.* Draft March 1995. Prepared for Central Hudson Gas & Electric Corporation, Poughkeepsie, New York.

[16] Weisberg, S.B., Burton, W.H., Jacobs, F. and Ross, E.A. (1987) Reductions in ichthyoplankton entrainment with fine-mesh, wedge-wire screens. *North American Journal of Fisheries Management* 7, 386–393.

[17] Taft, E.P., Hofmann, P., Eisele, P.J. and Horst, T. (1976) An experimental approach to the design of systems for alleviating fish impingement at existing and proposed power plant intake structures. *Third National Workshop on Entrainment and Impingement,* 343–365.

[18] McLaren, J.B. and Tuttle, L.R. (1999) Fish survival on fine mesh travelling screens. In: *EPRI, Power Generation Impacts on Aquatic Resources.* Conference Proceedings, 12–15 April.

[19] Fletcher, R.I. (1992) *In-House Trials of Fine Mesh Screening, Progress Report No. 2.* Prepared for Consolidated Edison Company of New York, New York.

[20] LMS (Lawler, Matusky & Skelly Engineers) (1996) *Effectiveness Evaluation of a Fine Mesh Barrier Net Located at the Cooling Water Intake of Bowline Point Generating Station.* Report prepared for Orange and Rockland Utilities Inc., January 1996.

[21] Taft, E.P. (2000) Fish protection technologies: a status report. *Environmental Science & Policy* 3, 349–359.

[22] LMS (Lawler, Matusky & Skelly Engineers) (2000) *Lovett Generating Station: Gunderboom Evaluation Program 1999.*

[23] Turnpenny, A.W.H. (1988) *The Exclusion of Salmonid Fish from Water Intakes.* CEGB report RD/L/3371/R88.

[24] Parker, G.C.C. (1977) Study of factors affecting sea water temperatures at Sizewell power station: an appraisal of the 1975 Sizewell hydrographic survey data. CERL laboratory report. Central Electricity Generating Board.

[25] Langford, T.E.L. (1990) *Ecological Effects of Thermal Discharges.* Elsevier, London.

[26] Eppley, R.W., Renger, E.H. and Williams, P.M. (1976) Chlorine reactions with sea-water constituents and the inhibition of photosynthesis of natural marine phytoplankton. *Estuarine and Coastal Marine Science* 4(2), 147–161.

[27] Smith, R.A., Brooks, A.S. and Jensen, L.D. (1974) Effects of condenser entrainment on algal photosynthesis at mid-Atlantic power plants. In: Jensen, L.D. (ed.) *Entrainment and Intake Screening. Proceeding of the 2nd Entrainment and Screening Workshop.* Rep 15. Edison Electric Institute, New York, pp. 113–122.

[28] Brylinski, J.M. (1981) Report on the presence of *Acartia tonsa* Dana (Copepoda) in the harbour of Dunkirk (France) and its geographical distribution in Europe. *Journal of Plankton Research* 3(2), 255–260.

[29] Bamber, R.N. and Seaby, R.M.H. (2004) The effects of power station entrainment passage on three species of marine planktonic crustacean, *Acartia tonsa* (Copepoda), *Crangon crangon* (Decopoda) and *Homarus gammarus* (Decapoda). *Marine Environmental Research* 57, 281–294.

[30] Bull, H.O. (1936) Studies on conditioned responses in fishes. Part VII. Temperature perception in teleosts. *Journal of the Marine Biological Association of the United Kingdom* 21(1), 1–27.

[31] Turnpenny, A.W.H. and Bamber, R.N. (1983) The critical swimming speed of the sand smelt (*Atherina presbyter* Cuvier) in relation to capture at a power station cooling water intake. *Journal of Fish Biology* 23(1), 65–73.

[32] Beach, M.H. (1984) *Fish Pass Design-Criteria for the Design and Approval of Fish Passes and Other Structures to Facilitate the Passage of Migratory Fish in Rivers.* Ministry of Agriculture, Fisheries and Food – Directorate of Fisheries Research.

[33] Zhou, Y. (1982) The swimming behaviour of fish in towed gears: a reexamination of the principles. *Department of Agriculture and Fisheries for Scotland* Working Paper 4, 1–55.

[34] Rose, A.H. (ed.) (1967) *Thermobiology.* Academic Press, London.

[35] Hokanson, K.E.F. (1977) Temperature requirements of some percids and adaptations to the seasonal temperature cycle. *Journal of the Fisheries Research Board of Canada* 34, 1524–1550.

[36] Cherry, D.S., Dickson, K.L. and Cairns, J. Jr (2011) Temperatures selected and avoided by fish at various acclimation temperatures. *Journal of the Fisheries Research Board of Canada* 32, 485–491.

[37] Snyder, G.R. (1969) Heat and anadromous fishes – discussion. In: Krenkel, P.A. and Parker, F.L. (eds) *Engineering Aspects of Thermal Pollution: Proceedings.* Vanderbilt University Press, Nashville, Tennessee, USA, pp. 318–337.

[38] Coughlan, J. (1970) *Preliminary observations on fish living at elevated temperatures: Part 1 – salmon (Salmo salar) in seawater.* Central Electricity Generating Board Report No. RD/L/N50/70.

[39] Coutant, C.C. (1975) Temperature selection by fish. A factor in power plant impact assessments. In: *Environmental Effects of Cooling Systems at Nuclear Power Plants. Proceedings of Symposium, Oslo, August 1974.* STI/PUB/378. International Atomic Energy Agency, Vienna, pp. 575–599.

[40] Cherry, D.S., Dickson, K.L. and Cairns, J. Jr (2011) Temperatures selected and avoided by fish at various acclimation temperatures. *Journal of the Fisheries Research Board of Canada* 32, 485–491.

[41] Jobling, M. (1981) Temperature tolerance and the final preferendum: rapid methods for the assessment of optimum growth temperatures. *Journal of Fish Biology* 19(4), 439–455.

[42] Langford, T.E.L. (1990) *Ecological Effects of Thermal Discharges.* Elsevier, London.

[43] Barnett, P.R.O. (1971) Some changes in intertidal sand communities due to thermal pollution. *Proceedings of the Royal Society B: Biological Sciences* 177(1048), 353–364.

[44] Bamber, R.N. and Spencer, J.F. (1984) The benthos of a coastal power station discharge canal. *Journal of the Marine Biological Association of the United Kingdom* 64(3), 603–623.

[45] Bamber, R.N. (1990) Power station thermal effluents and marine crustaceans. *Journal of Thermal Biology* 15(1), 91–96.

[46] Davis and Coughlan (1983) A model for predicting chlorine concentration within marine cooling circuits and its dissipation at outfalls. In: Jolley, R.L., Brungs, W.A. and Cotruvo, J.A. (eds) *Water Chlorination: Environmental Impact and Health Effects*, Volume 4. Ann Arbor Science, Ann Arbor, Michigan, USA, pp. 347–357.

6 Nuclear Generation

Nuclear power stations use nuclear power to generate the heat used to make the steam that powers the turbines that generate electricity. Therefore, they potentially have all the environmental issues discussed for steam turbine plants in Chapter 5 (this volume). When possible, direct, once-through cooling is preferred, possibly because there is a risk that radioactive material released from the reactor might concentrate in the cooling towers. The ecological impacts of direct-cooling, in particular impingement and entrainment of aquatic organisms, is discussed in Chapter 5 (this volume).

The world's first commercial nuclear power station was opened in England at Calder Hall, Cumbria, in 1956. Bradwell Nuclear Power Station in Essex opened several years later.

There are many designs of nuclear reactor and the most common is the pressurized water reactor. Figure 6.2 shows the basic design and Fig. 6.3 shows a typical example. Heat is taken from the reactor by the primary coolant, which is then transferred via a heat exchanger in the steam generator to the secondary circuit, which drives the turbines. Most nuclear reactors use uranium as fuel.

Installed Capacity

Most nuclear power stations are operated in Europe, Northern America and Asia. France is the largest proponent of nuclear power with about 72% of its electricity produced by nuclear plants. At present China is the only country with a growing nuclear power station fleet with about 30 reactors under construction. In many countries, nuclear power generation is set to decline as older plants are decommissioned. This decline

Fig. 6.1. View from the foreshore towards Hinkley Point A and B Nuclear Power Stations. The advanced gas reactor design of Hinkley B is in the foreground and the older Magnox reactor of Hinkley A behind. (Photo courtesy of Pisces Conservation Ltd, from CEGB archive.)

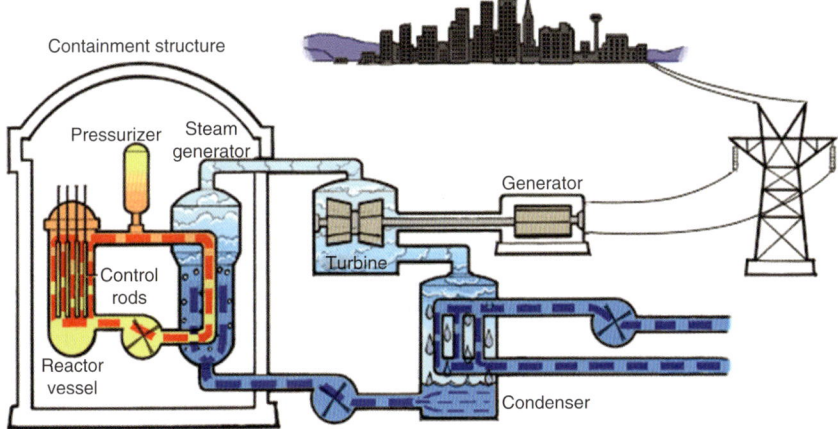

Fig. 6.2. Diagram of a pressurized water reactor. (From US NRC, public domain via Wikimedia Commons.)

is linked to both economic and safety concerns. The Fukushima Daiichi nuclear disaster in 2010 was a major turning point; the world had just about recovered from the shock of Chernobyl when the tsunami struck Japan. In Europe and North America there is little enthusiasm for nuclear

Fig. 6.3. Sizewell B Nuclear Power Station, Suffolk, UK. The only pressurized water reactor built in the UK. (Photo courtesy of Pisces Conservation Ltd.)

generation. Italy, Belgium, Germany, Spain and Switzerland have or will phase-out nuclear power. Global nuclear electricity generation in 2012 was at its lowest level since 1999. It is notable that the USA, with by far the largest number of reactors (see Table 6.1) has effectively stopped building or planning new nuclear plants.

In addition to the risks from accidental failure, there is the risk of terrorist attack. This has resulted in an ever-greater security cost. The 9/11 attack on New York alerted the USA to the fact that a plane could have hit the Indian Point Nuclear Power Plant on the Hudson river which was simply not designed to withstand the impact of a commercial airliner. The economic and social cost of such an attack would be incalculable. Many who consider the normal operation of a nuclear plant to be very safe have now come to question the safety of such plants when faced with terrorist hate and irrationality.

Nuclear power proponents, such as EDF in France, point to the low carbon footprint of nuclear power generation. It is the only selling point they have left. When I was young they were claimed to be the cheapest method of generation. This turned out to be untrue. They were claimed to be so safe a major accident would only occur once in many thousands of years. Another false assertion. We were told that decommissioning was already fully priced in and the money available. Another over-optimistic claim. Nuclear plants also have a terrible record of not being built within budget and on time.

Table 6.1. Total nuclear power installed capacity by country in 2016. (From IAEA, 2017.[1])

Country	Number of operated reactors	CapacityNet-total (MWe)	Generated electricity (GWh)	%-share of domestic generation
Argentina	3	1,632	7,677.36	5.6
Armenia	1	375	2,194.85	31.4
Belgium	7	5,913	41,430.45	51.7
Brazil	2	1,884	14,970.46	2.9
Bulgaria	2	1,926	15,083.45	35
Canada	19	13,554	95,650.19	15.6
China Mainland	36	31,384	197,829.04	3.6
Czech Republic	6	3,930	22,729.87	29.4
Finland	4	2,764	22,280.1	33.7
France	58	63,130	386,452.88	72.3
Germany	8	10,799	80,069.61	13.1
Hungary	4	1,889	15,183.01	51.3
India	22	6,240	35,006.83	3.4
Iran	1	915	5,923.97	2.1
Japan	43	40,290	17,537.14	2.2
Korea (South)	25	23,077	154,306.65	30.3
Mexico	2	1,552	10,272.29	6.2
Netherlands	1	482	3,749.81	3.4
Pakistan	4	1,005	5,438.9	4.4
Romania	2	1,306	10,388.20	17.1
Russia	37	26,528	184,054.09	17.1
Slovakia	4	1,814	13,733.35	54.1
Slovenia	1	688	5,431.27	35.2
South Africa	2	1,860	15,209.47	6.6
Spain	7	7,121	56,102.44	21.4
Sweden	10	9,740	60,647.40	40.0
Switzerland	5	3,333	20,303.12	34.4
Taiwan	6	5,052	30,461.09	13.7
Ukraine	15	13,107	76,077.79	52.3
United Kingdom	15	8,918	65,148.98	20.4
USA	100	100,351	804,872.94	19.7
World total	452	392,553 MWe	2,476 TWh	10.9

Environmental Issues

The main operational impacts of nuclear plants are the same as those for other plants using steam turbines (see Chapter 5, this volume).

Nuclear-specific safety issues are beyond the scope of a book of this size and are a specialized field. However, it is important to identify the main environmental issues. The key environmental concerns specific to nuclear plants relate to: (i) the production of nuclear fuel; (ii) the safe operation of the reactor and the release of radioactive material; (iii) the effects of nuclear accidents and terrorist attack; (iv) environmental issues relating to fuel reprocessing; (v) decommissioning; and (vi) long-term storage of nuclear waste.

Uranium mining

Uranium production is concentrated in three countries: Kazakhstan, Canada and Australia. The mining and milling of uranium has the environmental problems typical of mining activities, including surface and ground water contamination and the disposal and safe keeping of tailings.

The release of radioactive elements

The release of radioactive elements is a perpetual concern and always raised by objectors to nuclear power plants. A typical area of concern is the gradual release of tritium, which can be difficult to contain.

Containment failure

This type of accident, which is normally claimed to have a vanishingly small probability of occurring, has unfortunately occurred twice in the last 40 years: at Chernobyl in 1986 and at Fukushima in 2011. In both cases the economic and social effects were catastrophic and are even now far from over.

Events at Chernobyl are the worst nuclear power plant disaster to date. The eventual death toll is unknown, possibly in the thousands, but far from clear. Radioactive contamination was spread across Europe, even reaching North Wales where it contaminated farm stock via the soil and grass. The only positive outcome of this event is that the Chernobyl Exclusion Zone, an area of 490 km², has been designated a nature reserve. The extent of the Fukushima disaster is yet to be fully realized, because the damaged plant is not yet fully contained. It is clear that the economic cost will dwarf that of Chernobyl and it is even possible the release of radioactive material will be of a similar magnitude.

Fuel reprocessing

Nuclear fuel reprocessing is used to separate out useful elements from spent nuclear fuel. It is a contentious area with a huge and angry literature. Unfortunately, commercial nuclear electrical generation becomes mixed with military needs for plutonium and other elements.

Decommissioning

A surprisingly small number of commercial power-generating reactors have been fully decommissioned to date. It sometimes seems that some older plants are kept in commission, even if not very profitable, because the decommissioning costs are so high. It is becoming clear that the process is considerably more time-consuming and costly than originally envisaged.

Long-term storage of nuclear waste

The seemingly perpetual arguments about high-level radioactive waste disposal is a major issue for the nuclear power industry. While most engineers and scientists favour deep geological burial, either in a mine or a deep borehole, no government has succeeded in opening such a repository for civilian high-level nuclear waste. This might be about to change, because Finland is constructing the Onkalo spent nuclear fuel repository.

Spent nuclear fuel rods are classified as high-level nuclear waste (HLW). Fuel rods contain fission products and transuranic elements generated in the reactor core. Spent fuel is highly radioactive and often hot. The amount of HLW worldwide is currently increasing by about 12,000 metric tons every year and a 1000-MW nuclear power plant produces about 27 tonnes of spent nuclear fuel per year.

It is a continuing cause for concern that the nuclear power industry can keep arguing for more nuclear power reactor construction without any firm plans where or how the waste is to be stored long-term. Given the extremely long half-lives of some radioactive waste, it is clear that this is a problem which we continue to add to and which will need to be addressed by mankind far into the future. Plutonium 239 has a half-life of 24,100 years. The Roman republic and empire together only lasted about 2000 years. We are creating dangerous waste with a half-life much greater than that of our societies and political systems. We are already having to address the security of nuclear materials in regions suffering civil strife and political breakdown. The general response of the nuclear industry is to point out the relatively small volume of the waste when compared to a large building or stadium. Efficient states can certainly secure their waste, but, judging by history, their collapse must be viewed as inevitable.

Note

[1] IAEA (2017) Nuclear Share of Electricity Generation in 2016. Available at: https://www. iaea.org/PRIS/WorldStatistics/NuclearShareofElectricityGeneration.aspx (accessed 19 April 2017).

7 Coal and Oil-Fired Power Plants

Both coal- and oil-fired power stations use steam turbines to generate electricity and therefore need to address all the issues covered in Chapter 5 (this volume). A schematic diagram of a typical coal plant is shown in Fig. 7.2. Large-scale oil-fired generation is no longer common and needs little detailed consideration here. Many of its polluting characteristics, such as sulphur dioxide (SO_2) and nitrogen oxides (NO_x) emissions, mirror those of coal. Coal is the dominant energy source for electrical power generation on a global basis, producing about 41% of total production, and it will remain a major energy source for the foreseeable future.

The nature of coal and the waste generated by its combustion create a number of issues that, while not unique to coal combustion, are particularly problematical for coal plants. It is the environmental problems associated with coal handling and combustion that are considered in particular detail in this chapter.

Coal Mining and the Environment

First, it should be noted that mining is a dangerous activity and miners have been killed and injured throughout recorded history. This is illustrated by the fact that even in a safety-conscious country such as the USA, an average of 23 coal miners per year died in the decade 2007–2016 (US Department of Labor, https://arlweb.msha.gov/stats/centurystats/coalstats.asp). There is also a long history of occupational illnesses such as pneumoconiosis. The illness and death rates accepted in the coal-mining industry would cause outrage if they were to occur at facilities handling nuclear material. While the focus of the present work is the environmental impact linked to electrical power generation, it is essential to recognize that the mining of coal also creates considerable environmental impacts.

Fig. 7.1. The 2000-MW Kingsnorth Power Station, Thames Estuary, England. The image shows the coal-handling jetty and coal conveyers. This power station was designed to burn both oil and coal. (Photo courtesy of Pisces Conservation Ltd, from CEGB archive.)

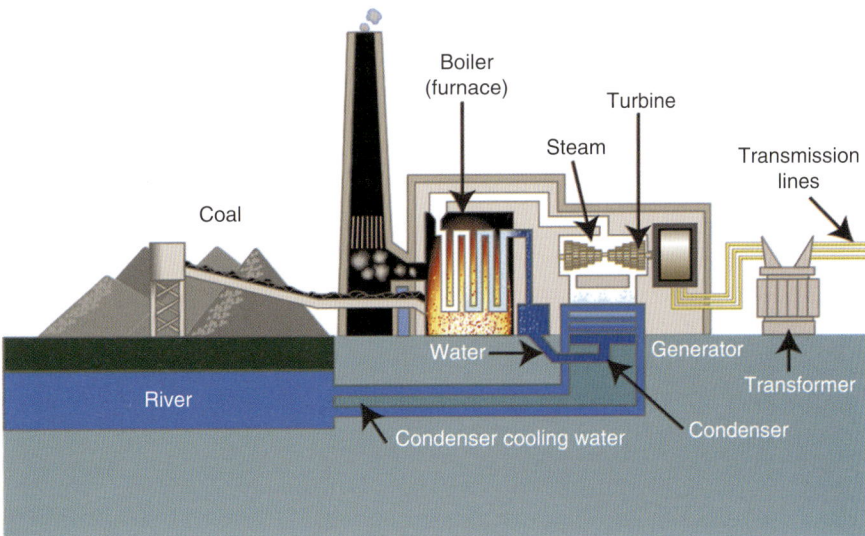

Fig. 7.2. Diagrammatic layout of a typical coal-fired power station. (From Tennessee Valley Authority, public domain via Wikimedia Commons.)

The key areas of concern are:

- Land disturbance and degradation (see Fig. 7.3).
- Mining subsidence.
- Water pollution and management.
- Dust and noise.

Land disturbance and degradation

Both open-cast and deep-shaft mining create considerable land disturbance and habitat loss. Strip mining, which produces much of the coal used in power stations, destroys the existing vegetation and top soil, displaces or kills wildlife, causes a loss of previous land-based activities and changes the topography of the area mined. Scenic, archaeological, paleontological and historical features of the landscape may be lost. Spoil heaps alter the topography, destroy habitat and can create dangerous structures (see Figs 7.3 and 7.4).

Mining subsidence

Mine subsidence is where the ground level falls as a result of mining. This is a common problem in historically heavily mined areas. For example, in the German Saar region, a historical centre for mining, a mine collapse in 2008 is believed to have created a minor earthquake measuring 4.0 on the Richter scale.

Fig. 7.3. An open-cast coal mine at Dhanbad, India. (Published under a CC BY 2.0 license via Wikimedia Commons.)

Fig. 7.4. Open-cast coal-mining in Wyoming, USA. (Photo courtesy of US Federal Government, public domain.)

Water pollution and management

Mining interferes with the natural flows and storage of water. A particular problem is the surface disposal of water pumped from mine workings. Acid mine drainage (AMD) is metal-rich water formed from the chemical reaction between water and rocks containing sulphur-bearing minerals. It is a common problem and there are many examples of deliberate and accidental contamination of surface waters resulting in great loss to aquatic life. It can become a particular problem in abandoned mine workings which gradually become flooded and then release toxic waters. Acid mine drainage has huge costs for government and the mining industry. In Australia, Canada and the USA, total costs of AMD are hundreds of millions of dollars per year.

Dust and noise

The movement of coal generates considerable dust and noise pollution. Coal is transported by ship, rail, conveyers and road and all cause notable local dust and noise pollution (see Figs 7.2 and 7.5 for a typical example of a power station coal-handling system). Coal stock heaps at power stations need to be constantly moved and sorted, in part to avoid the build-up of heat and spontaneous combustion. Coal dust contains toxic substances, such as sulphur dioxide, hydrogen chloride, mercury, arsenic, cadmium and other heavy metals. While coal particles can occur naturally in waters and result

Fig. 7.5. A typical coal-handling system at a power station. The image shows the conveyor system from the jetty to the onshore coal stock heaps. From the stock heap the coal is transported by another conveyer into the mill. Kilroot Power Station, Northern Ireland. (Photo courtesy of R. Somes, Pisces Conservation Ltd.)

in no ill effects, the pollution of surface waters by coal dust can result in a reduction of dissolved oxygen and a loss in aquatic flora and fauna.

Radiation Release

Coal contains small amounts of radioactive minerals. In a survey of US coal, the US Geological Survey concluded that in the majority of samples concentrations of uranium and thorium each range from about 1 to 4 parts per million (ppm). These concentrations are similar to those found in a range of common rocks and soils. Concern has been generated because the nuclear industry and its lobbyists are keen to point out that coal combustion releases greater quantities of radioactive material than nuclear power plants. Such arguments have become less compelling and more muted after the Fukushima disaster (see Chapter 6, this volume).

Oil Drilling, Fracking and Transportation Impacts

Impacts relating to oil pollution are well known and widely discussed. There is an immense literature on this subject and there is no point reviewing it here. The small number of large oil-fired power plants now in operation consume a negligible proportion of the oil produced and transported. There is, however, a large amount of refined oil product used in small-scale diesel and gas turbine plants (see Chapter 14, this volume). Fracking impacts are considered in Chapter 8 (this volume).

Ash Disposal and Use

Fly ash

Fly ash or pulverized fuel ash produced by the combustion of coal dust is composed of fine particles removed from the flue gas using electrostatic precipitators and filters. The chemical composition varies with the type of coal, but all fly ash has a high content of silicon dioxide (SiO_2) (both amorphous and crystalline), aluminium oxide (Al_2O_3) and calcium oxide (CaO), which are the main mineral compounds in coal-bearing rock strata. There are almost inevitably also trace concentrations of arsenic, beryllium, boron, cadmium, chromium, hexavalent chromium, cobalt, lead, manganese, mercury, molybdenum, selenium, strontium, thallium and vanadium. As shown in Figs 7.6 and 7.7, ash storage takes up a considerable amount of land on coal-fired power station sites.

Fig. 7.6. Kilroot Power Station, Northern Ireland, UK. The ash lagoons are in the foreground in front of the coal storage area. (Photo courtesy of R. Somes, Pisces Conservation Ltd.)

Fig. 7.7. The ash dump at Kilroot Power Station, Northern Ireland. (Photo courtesy of R. Somcs, Pisces Conservation Ltd.)

Fly ash presents a major waste-disposal problem for coal-fired power plant operators. It is not generally viewed as hazardous to man or the environment. In 2000, the US Environmental Protection Agency (EPA) said that coal fly ash did not need to be regulated as a hazardous waste.[1] However, its storage and disposal can have highly detrimental environmental effects.

Worldwide, more than 65% of fly ash produced from coal power stations is disposed of in landfills and ash ponds. The industrial use of fly ash has become increasingly attractive as landfill becomes more expensive and limited, and dumping at sea is no longer permitted. As reported at Coal Ash Facts (http://www.coalashfacts.org/), in 2007, the USA produced 131 million tons of coal combustion products, of which 43% were used beneficially and the remainder, nearly 75 million tons, disposed of to landfill. (An extensive range of publications on the use of fly ash are available from The American Coal Ash Association at https://www.acaa-usa.org/Publications/Free-Publications.)

The main industrial uses include:

- Concrete production, as a substitute material for Portland cement and sand.
- Road construction fill material for embankments and sub-bases.
- Grout and flowable fill production.

- Waste stabilization and solidification.
- Cement clinkers production.
- Mine reclamation.
- Stabilization of soft soils.
- Aggregate substitute in brick production.
- Mineral filler in asphaltic concrete.

There are also a multitude of uses for small amounts in many manufactured products.

Environmental impacts of fly ash

A key concern is the leaching of contaminants, heavy metals in particular, from ash lagoons and dumps. The US National Academy of Sciences noted in 2006 that 'the presence of high contaminant levels in many CCR (coal combustion residue) leachates may create human health and ecological concerns'.[2]

Fly ash is stored wet to reduce dust levels. Impoundments at large power station sites are frequently extensive and result in habitat degradation. However, old ash lagoons can hold abundant aquatic life. There have been failures in dams and bunds that resulted in the rapid release of large quantities of ash and contaminated water. For example, in December 2008, the collapse of an embankment retaining fly ash at the Tennessee Valley Authority's Kingston Fossil Plant caused a release of 5.4 million cubic yards of coal fly ash. In 2014, between 50,000 and 82,000 tons of ash and 27 million gallons (100,000 cubic m) of contaminated water spilled into the Dan River near Eden, NC from a closed North Carolina coal-fired power plant (see *Los Angeles Times*: http://www.latimes.com/nation/la-na-coal-ash-20140205-story.html#axzz2sUqN4ngB).

New regulations published in the *Federal Register* on 19 December 2015 stipulate a comprehensive set of rules and guidelines for safe disposal and storage of fly ash. These are designed to prevent pond failures and protect groundwater and include an enhanced inspection, record keeping and monitoring regime. Procedures for closure are also included and include capping, liners and the dewatering of ponds.

Bottom ash

In addition to fly ash, coal combustion also generates bottom ash. The quantities are smaller and thus generally less problematical. However, boiler bottom ash can hold high levels of heavy metals and toxic compounds.

Carbon Dioxide Emissions

Coal- and oil-fired power plants are among the world's major sources of atmospheric CO_2. The amount of CO_2 produced per unit of electricity generation varies with power plant design and efficiency. A typical 600 MW coal plant generates in the region of 3.5 million tons of CO_2 per year (assuming a capacity factor of 69%; heat rate of 10,415; CO_2 emissions rate of 206 pounds of CO_2/million Btu). Coal combustion is generally more carbon in-tensive than burning natural gas or petroleum for electricity, which is why CO_2 emissions are a particular problem for coal plants. To illustrate the point, for the USA in 2015, coal accounted for about 70% of CO_2 emis-sions from power plants, but contributed only about 34% of the electricity generated.

In 2009, total CO_2 emissions from power plants throughout the world were estimated to be approximately 1×10^{10} tonnes per annum. This compares with a slightly lower 9.53×10^9 tonnes in 2004. It is likely that annual production has increased since 2004 primarily because of increased use of coal-powered plants in China, India and Germany. The distribution of CO_2 emissions by country is shown in Fig. 7.8 for the 50 largest emitters. China and the USA dominate CO_2 production, and the top six emitters, China, USA, India, Russia, Japan and Germany, produced about 66% of the total.

Total CO_2 emissions are likely to continue to increase, primarily be-cause of developments in China and India. The USA is gradually reducing CO_2 production as coal plants are closed. In 2011, coal plants in the USA emitted a total of 1.7 billion tons of CO_2. Major declines in coal use have also occurred in other western countries. For example, Fig. 7.9 shows that within the UK the amount of power generated from coal combustion has greatly declined since 1990. Monthly UK power output from coal was zero for the first time in more than 150 years in March 2016. In April 2017, coal supplied 2–3% of UK electricity, down from around 45% in April 2012. Most remarkably, the UK generated more electricity from wind than from coal in 2016. European countries in particular have greatly reduced CO_2 emissions from coal over the last 15 years because it has been re-placed by solar, wind and gas generation in particular (see Fig. 7.8 for CO_2 emissions by country). It is likely that a similar path will be followed by the USA over the next 10 years, especially given the availability of locally produced gas.

Why is CO_2 considered detrimental to the environment?

Carbon dioxide at an appropriate atmospheric concentration is essential to life on earth. CO_2 helps to keep the planet at a suitable temperature for life and also acts as the carbon source used by photosynthetic plants to produce sugars. Environmental concerns and arguments centre on the fact that coal combustion has contributed to a general increase in CO_2 levels in the atmosphere observed since the start of the industrial revolution. It is generally believed that this increase has increased average global temper-atures and CO_2 is a major cause of global warming. Further, CO_2 can also cause ocean acidification.

The mechanism by which CO_2 traps heat in the atmosphere is com-monly referred to as the 'greenhouse effect'. CO_2 allows incoming solar

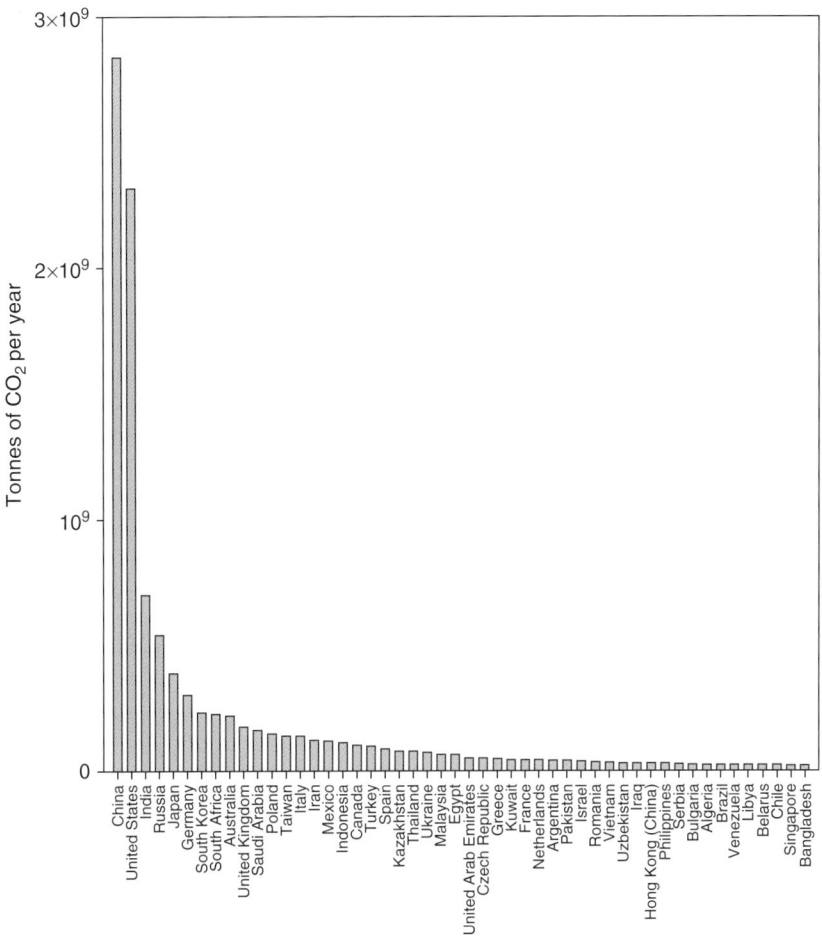

Fig. 7.8. Total annual CO_2 emissions from the combustion of coal in power stations by country for the year 2009. Only the top 50 emitters are plotted. (Data from the Carbon Monitoring for Action website, http://carma.org/.)

radiation from the sun to pass and warm the earth's surface. However, when the surface, in turn, emits a portion of this energy skywards at an increased wavelength, some of this thermal radiation is absorbed and re-radiated by the atmosphere's CO_2 molecules back towards the earth's surface, providing an additional source of heat energy. While the physics of this process is not in doubt, there are considerably greater uncertainties about the actual relationship between CO_2 concentration and the amount of warming actually produced. In part, this is because of the influence of life which responds to changes in CO_2 concentration and temperature and also other factors such as cloud formation and the presence of other greenhouse gases.

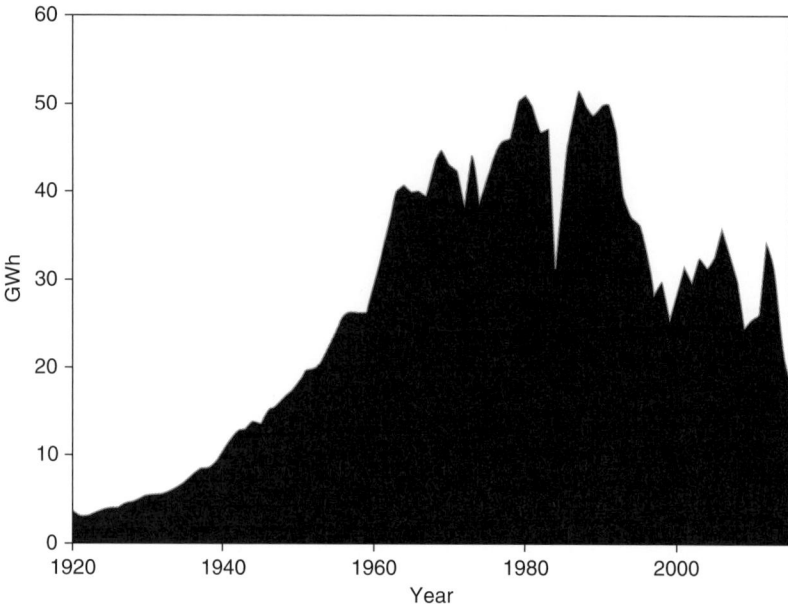

Fig. 7.9. The change in the total annual electrical power generated using coal within the United Kingdom between 1920 and 2015. (From P.A. Henderson, data from the Carbon Monitoring for Action website http://carma.org/.)

Can CO$_2$ pollution be reduced?

The latest coal-fired power stations are close to maximum thermodynamic efficacy, so the only possible pathway to produce appreciable reductions would be to stop the CO$_2$ entering the atmosphere.

Carbon capture and storage or sequestration
Carbon capture and storage (CCS) refers to the capture of CO$_2$ from large point sources, such as power plants, transporting it to a storage site and depositing it where it will not enter the atmosphere (Fig. 7.10). Storage is typically an underground geological formation. While CO$_2$ has been injected into geological formations for the past several decades for various purposes, including enhanced oil recovery, the long-term storage of CO$_2$ is still an experimental concept and not widely applied. The first commercial example was the Weyburn-Midale Carbon Dioxide Project in 2000. Other examples include SaskPower's Boundary Dam and Mississippi Power's Kemper Project. A 90-MW pilot-scale demonstration CCS power plant began operation in September 2008 at the German Schwarze Pumpe power station to test the technological feasibility and economic efficiency. CCS applied to a modern conventional power plant could reduce CO$_2$ emissions to the atmosphere by approximately 80–90% compared with a plant without CCS. A general problem is that long-term predictions as to

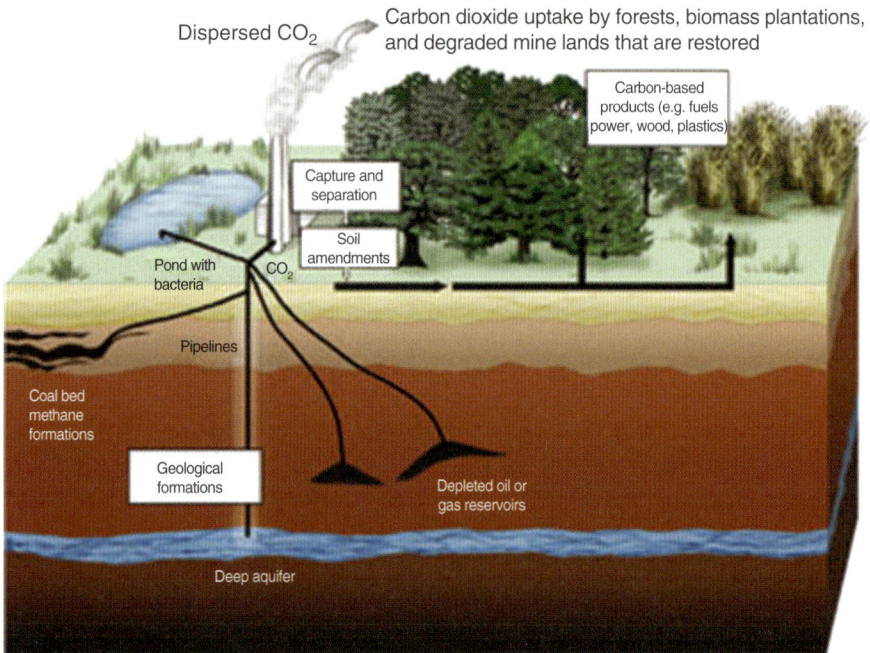

Fig. 7.10. Schematic showing both terrestrial and geological sequestration of CO_2 emissions from a coal-fired plant. (Rendering by LeJean Hardin and Jamie Payne. From http://www.ornl.gov/info/ornlreview/v33_2_00/research.htm, published under a CC BY-SA 3.0 license via Wikimedia Commons.)

the stability of submarine or underground storage are uncertain, and there is still the risk that CO_2 might leak into the atmosphere.

Carbon dioxide can be captured from the power plant flue gas using carbon scrubbing, membrane gas separation or adsorption technologies. The German utility RWE operates a pilot-scale CO_2 scrubber at the lignite-fired Niederaußem power station.

CCS is a commercially unproven technology with a high energy cost of between 10 and 40% of the energy produced by a power station. There would inevitably be an appreciable cost to the consumer. While costs are uncertain, it has been estimated that the cost of capturing and storing CO_2 is in the region of US$60 per ton, corresponding to an increase in electricity prices of about US6c per kWh (based on typical coal-fired power plant emissions of 0.97 kg CO_2 per kWh). This would double the typical US industrial electricity price.

Sulphur Dioxide and Acid Rain

The largest single source of SO_2 to the atmosphere is the burning of fossil fuels by power plants and manufacturing industries. Smaller sources include smelting and transportation, such as locomotives and ships that burn fuel with a high sulphur content. There are also significant natural sources such as volcanoes and the oxidation of dimethyl sulphide derived from ocean plankton.

Coal power plants are the largest single source of SO_2. A typical 600-MW coal plant without SO_2 reduction technology emits in the region of 14,100 tons of SO_2 per year. This is reduced to 7000 tons of SO_2 per year when fitted with emissions controls, including flue gas desulphurization scrubbers. Oil-fired plants can also emit considerable quantities of SO_2, depending on the type of oil burnt.

The amount of SO_2 generated by coal combustion varies greatly with the sulphur content of the coal. Low-sulphur coal, with a sulphur content of less than 0.5%, holds sulphur from the original plant material from which it was derived. High-sulphur coal often contains mineral sulphur largely as pyrites. This is believed to be derived from the infiltration of sea water. However, high-sulphur coals with a high organic sulphur content do occur. Pittsburgh 8 coal, for example, may contain 1.5% organic sulphur.

SO_2 is a precursor of atmospheric sulphuric acid (H_2SO_4) and sulphate aerosol. SO_2 is oxidized by hydroxyl radicals (OH) to form sulphuric acid, which then becomes aerosols through photochemical gas-to-particle conversion. SO_2 dissolves readily in suspended liquid droplets in the atmosphere.

Acid rain damages crops, forests and soils, and acidifies lakes and streams. The effects are particularly noted in regions with naturally acidic conditions, for example in areas with a granitic geology and lacking limestone.

Why is SO_2 detrimental to our environment?

Short-term exposures to SO_2 can harm the human respiratory system and make breathing difficult. Children, the elderly and those who suffer from asthma are particularly sensitive to the effects of SO_2.

SO_2 emissions that lead to high concentrations of SO_2 in the air generally also lead to the formation of other sulphur oxides (SOx). SOx can react with other compounds in the atmosphere to form small particles. These particles contribute to particulate matter (PM) pollution: particles may penetrate deeply into sensitive parts of the lungs and cause additional health problems.

Flue gas desulphurization

There is a long history of concern about SO_2 emissions. The first major flue gas desulphurization (FGD) plant was operated from 1931 at Battersea Power Station, London. This was followed by FGD systems at Swansea and Fulham Power Stations. These were abandoned during the Second World War and FGD units at power stations were only reintroduced in the 1970s after considerable resistance from the generating industry was overcome by growing concern about acid rain and human health effects.

There are a number of different processes available to remove SO_2 and many more have been proposed. The main approaches presently used are:

- Dry absorption.
- Limestone – gypsum.
- Wellman Lord regenerative.
- Seawater scrubbing.

There are two main methods for desulphurization of flue gases from coal-fired power stations: dry scrubbing and wet scrubbing. An alternative to removing sulphur from flue gases is to remove the sulphur before or during combustion. Hydrodesulphurization of fuel has been used for treating fuel oils before use. Fluidized bed combustion adds lime to the fuel during combustion. The lime reacts with the SO_2 to form sulphates, which become part of the ash. Fluidized bed combustion, once viewed as a potentially important development, has lost importance as the need to reduce carbon and hence coal use became a major environmental objective.

Costs of FGD

According to the US Energy Information Administration (EIA), the historical costs of adding flue gas desulphurization (FGD) equipment to remove sulphur dioxide are, in 2006 US dollars, $301/KW for a 300 MW plant, $230/KW for a 500 MW plant and $190/KW for a 700 MW plant. In 2016 the EPA gave the average installed capital cost as $288.27/KW.[3] FGD units are assumed to remove 95% of the SO_2.

Dry lime scrubbing

In dry scrubbing, slaked lime is injected directly into flue gas to remove SO_2 and Hydrochloric acid (HCl). There are two major dry processes: dry injection systems inject dry hydrated lime into the flue gas duct, and spray dryers inject an atomized lime slurry – the water in the lime slurry is evaporated by the hot gas. Both dry injection and spray dryers yield a dry final product, collected in particulate control devices. For power stations, dry scrubbing is used primarily with low-sulphur fuels.

Dry FGD systems have the following advantages when compared with wet limestone forced oxidation (LSFO):

- The capital cost is typically lower.
- Overall power consumption is lower.
- The by-product is dry and can be handled with conventional fly ash handling equipment.
- The by-product is stable for most landfill purposes and usually can be disposed with fly ash.
- The dry system is a less complicated process and has lower maintenance labour requirements.
- Pressure drop across a spray dryer absorber is lower.
- High chloride levels in the fuel improve, rather than hinder, SO_2 removal performance.

- Sulphur trioxide (SO_3), which condenses to sulfuric acid aerosols in an FGD system, is removed efficiently (>90%) with a dry FGD baghouse. (A baghouse is a filter made of fabric which is used in power stations to remove particulate air pollutants.)
- Flue gas following dry scrubbing is not saturated with water, which reduces or eliminates the visible moisture plume seen from the stack.
- There is no liquid waste that may require a wastewater treatment plant.

From an environmental perspective, there are two clear disadvantages compared with wet scrubbing.

- Dry FGD produces a by-product that has fewer beneficial uses and in greater quantities.
- The combined removal of fly ash and by-product solids in the particulate collection system precludes commercial sale of the fly ash.

Wet lime scrubbing

In wet lime scrubbing, lime is added to water and the resulting slurry is sprayed into a flue gas scrubber. In a typical system, the gas to be cleaned enters the bottom of a cylinder-like tower and flows upward through a shower of lime slurry. The SO_2 is absorbed into the spray and then precipitated as wet calcium sulphite. The sulphite can be converted to gypsum, a saleable by-product. Wet scrubbing treats high-sulphur fuels and some low-sulphur fuels where high-efficiency SO_2 removal is desirable. Wet scrubbing primarily uses magnesium-enhanced lime (containing 3–8% magnesium oxide) because it provides high alkalinity to enhance SO_2 removal and reduce scaling. The majority of installations at power plants are wet scrubbers.

While wet scrubbing can produce commercially useful gypsum used to produce plaster boards and other products, in practice the market is easily saturated and the value is too low to allow long-distance transportation.

ENVIRONMENTAL ISSUES LINKED TO WATER USE AND DISCHARGE FROM WET SCRUBBING PLANTS.   FGD plants consume water, principally by evaporation. Quantities are not great – possibly 250 m³/h for a large power station. However, in an arid region these amounts can be important. The discharge from a freshwater FGD plant would typically be released with the cooling water stream if direct cooled or, if cooling towers are used, together with the cooling tower purge water. The limestone-gypsum FGD process would typically produce a temperature increase of around 0.8°C in the temperature of the discharge stream. The pH of the cooled water discharge would be alkaline, but within the natural range of surface waters.

FGD discharges are likely to have a chemical oxygen demand (COD) of about 4.24 mg/l. Any reduction in oxygen levels will first affect fish, as they typically consume more oxygen than invertebrates and depend on dissolved oxygen.

The discharge typically contains a wide range of elements and compounds at concentrations above those naturally occurring in the river water. As an example of what can be present, Table 7.1 presents the predicted concentrations for a proposed FGD wet scrubbing discharge at Didcot, England. The most abundant components in the discharge are chloride, sulphate and nitrates. The chloride content is possibly high enough close to the outfall to lead to the exclusion of any freshwater life not adapted to slightly brackish conditions. This will include many larval insects. The total annual input to the river of nitrates was estimated at up to 8000 tonnes. Such an input will contribute to the eutrophication of the river. Eutrophication is the unnatural enrichment of surface waters by nutrients, particularly nitrogen and phosphorus compounds.

When considering the impacts of chemical pollutants, particularly heavy metals, it is not only the concentration but also the total loading that must be considered. A particular concern with some elements is their tendency to accumulate in organic-rich sediments and enter the food chain. In the case of the Didcot discharge, arsenic, cadmium and mercury discharges were predicted to exceed significant load levels when calculated up from µg/l to kg released per year (Table 7.2). Significant load limits apply to the total amount released, rather than the resultant concentration on mixing with background flows. In addition, appreciable annual inputs of chromium, lead, selenium, vanadium and copper would also occur.

Table 7.1. Discharge concentrations for a proposed FGD discharge at Didcot Power Station on the Thames, England. The FGD plant was not built at this site. (Data supplied by Pisces Conservation Ltd.)

Material	Mean discharge concentration (µg/l)
Antimony	14.92
Arsenic	16.74
Boron	2,690.96
Cadmium	1.14
Chloride	906,945.3
Chromium	32.26
COD	10,923.4
Copper	67.58
Fluoride	858.36
Lead	15.97
Manganese	129.12
Mercury	0.764
Nickel	15.18
Nitrate	170,480.94
Selenium	29.86
Sulphate	281,359.2
Suspended solids	34,777.24
Vanadium	19.21
Zinc	74.96

Table 7.2. Background and additional flux of trace metals and other determinands in FGD effluent after mixing for a proposed FGD discharge at Didcot Power Station on the Thames, England. (Data supplied by Pisces Conservation Ltd.)

	Total released (kg/year)	Threshold of significant load(kg/year)	Ratio of exceedance
Arsenic	209.7	20	10.5
Cadmium	11.0	5	2.2
Mercury	11.0	0.5	22.1

Seawater scrubbing

This approach can only be taken at estuarine and coastal power plants. A proportion of the power station seawater cooling flow is used to absorb SO_2 and generate sulphate ions. Sea water is naturally high in sulphates and the water flow is discharged to the sea, where the sulphate is dispersed. There are a number of potential environmental impacts of seawater scrubbing linked to the changed chemistry of the cooling water discharge stream.

Reduced pH

The most important characteristic of effluent seawater from seawater FGD wash process plants is the increased acidity. Even when partially neutralized by mixing with the remainder of the cooling water stream, the effluent may be pH 6.0 at the discharge point. Fully marine seawater is alkaline, buffered by the equilibrium of CO_2 and carbonate that it contains; its natural pH levels generally range between 7.6 and 8.4. The rarity of naturally occurring reduced pH seawater means that there has been comparatively little research on its effects on marine organisms (Batten and Bamber, 1996),[4] although the power industry's interest in FGD has provided some impetus over the last 30 years. Despite the comparative lack of experimentation on acid seawater, the effects of low-pH water on marine organisms have been reasonably well-documented, chiefly by Bamber (1985, 1987, 1990)[5, 6, 7] and associates. Tables 7.3 and 7.4 briefly summarize the major physiological effects reported at a range of pH values. Table 7.5 notes threshold pH values for lethal and sub-lethal effects, while Table 7.6 summarizes other responses of bivalve molluscs to low pH. Data are taken from Bamber (1985, 1987, 1990), Batten and Bamber (1996) and Langford and Bamber (1984)[8].

Bamber (1990) concludes, in reference to his own and other studies: 'While some significant deleterious effects are found at pH 7.5 (surprisingly close to the lower limit of natural seawater), it is clear from these data and the present results that seawater at pH 7.0 is intolerable to bivalve molluscs.'

Langford and Bamber (1984), referring specifically to fish, summarize a range of effects, including avoidance behaviour beginning at pH 6.5, discoloration and eye damage at around pH 6, increased mortality at pH 5.5 and 100% mortality below pH 5. They state: 'No data on pH effects at

Table 7.3. Effects of lowered pH levels on various marine invertebrates (effects of pH < 5 not shown). (Data from Bamber, 1985, 1987, 1990[5,6,7] and Batten and Bamber, 1996.[4])

Organism	<5.5	<6.0	<6.3	<6.5	<7.0	<7.5
Carpet-shell clam, *Venerupis decussata*	Activity minimal (<pH 5.4)	Extreme shell fragility, reduction of activity, valve closure; hyper-extension of siphon tube	50% mortality within 1 month (larger specimens, <pH 6.3), evident shell corrosion (<pH 6.1)	50% mortality within 18 days (smaller specimens, pH 6.4)	Suppression of feeding activity and growth. Survival of young clams to reproductive age negligible	Shell dissolution
Oyster, *Crassostrea gigas*		Significant mortality		Abnormal behaviour, shell gaping, torpor (<pH 6.6).	Growth suppression, tissue weight loss, reduced shell size, reduced shell density, shell dissolution, suppression of feeding (<pH 7.2). Absence of byssal attachment (<pH 6.9). Increased mortality	
Oyster, *Ostrea edulis*				Abnormal behaviour, shell gaping, torpor (<pH 6.6).	Growth suppression, tissue weight loss, reduced shell size, shell dissolution, suppression of feeding (<pH 7.2). Significant mortality (pH 6.9 – young, pH 7.0, spat). Absence of byssal attachment (<pH 6.9)	

Continued

Table 7.3. Continued.

Organism	<5.5	<6.0	<6.3	<6.5	<7.0	<7.5
Mussel, *Mytilus edulis*				Abnormal behaviour, shell gaping, torpor (<pH 6.6). Significant mortality (pH 6.6) Absence of byssal attachment (<pH 6.6)	Growth suppression, tissue weight loss, reduced shell size, shell dissolution, suppression of feeding (<pH 7.2). Increased mortality	
Oyster, *Ostrea virginica*				Increased valve closure		
Shellfish, general				Abrupt changes in irritability and behaviour		
Ragworm, *Nereis virens*		Marked decline in tissue glycogen levels, indicating use of stored reserves required to maintain metabolism		Increased mortality, reduced dry weight, reduced burrowing. Decline in tissue glycogen levels. (All at 18°C) Depression of body weight (9°C and 18°C)		Reduced burrowing activity, at 18°C

Table 7.4. Effects of various lowered pH levels on fish (effects of pH levels <5 not shown). (Data from Langford and Bamber, 1984.[8])

Organism	≤5	<5.5	<6.0	<6.3	<6.5
Fish, general					Signs of avoidance behaviour
Brown trout, *Salmo trutta* – in 50% seawater				Discoloration and blackening of skin; sluggishness; irreversible blistering and opacity of eyes; increased mortality	
Bass, *Dicentrarchus labrax*	100% mortality in 17.2 h	50% mortality over 96 h		Discoloration and blackening; sluggishness	
Rock goby, *Gobius paganellus*		Sluggishness; increased mortality			
Plaice, *Pleuronectes platessa*	50–100% mortality	Increased mortality			
Mullet, *Chelon labrosus*	Loss of equilibrium; sluggishness; disorientation	Increased mortality	Avoidance behaviour		
Dab, *Limanda*	100% mortality	83% mortality			

higher temperatures or on combined stresses of pH, heat and heavy metals are available. Evidence from published data suggests that the combined stresses will be more harmful than low pH alone, though effects at a discharge site will depend on effluent dispersal and avoidance behaviour, as well as on the pH thresholds of the fish present'.

Macro contaminants
The following contaminants are potentially present in effluent water on a macro rather than trace scale:

- chloride;
- sulphate;
- phosphorous/phosphates;
- nitrates; and
- suspended sediment.

Table 7.5. Reported critical pH levels for lethal and sub-lethal effects. (Data from Bamber, 1985, 1987, 1990,[5, 6, 7] Batten and Bamber, 1996[4] and Langford and Bamber, 1984.[8])

Organism	Threshold of lethal effects	Threshold of sub-lethal effects
Nereis virens		≤6.5
Mytilus edulis	6.6	≤7
Ostrea edulis	6.9	≤7
Crassostrea gigas	6.0	≤7
Venerupis decussate		≤7
Dicentrarchus labrax	5.6	6.3
Gobius paganellus	5.0	5.85
Pleuronectes platessa	5.5	6.0
Limanda limanda	5.5	5.5
Chelon labrosus	5.17	5.17
Salmo trutta	6.0	6.0

Table 7.6. Other responses of bivalves to low pH reported in literature. (Adapted from Bamber, 1990.[7])

Species	Effect	Critical pH	Authority
M. edulis	Reduced gamete respiration	7.6	Akberali *et al.* (1985)[9]
Pintada fucata	Increased adult mortality	7.48	Kuwatani and Nishii (1969)[10]
	Increased weight loss attributed to shell dissolution	7.66	Kuwatani and Nishii (1969)[10]
Crassostrea virginica	Adult reduced pumping rate	7.0	Loosanoff and Tommers (1947)[11]
	Abnormal adult shell movement	7.0	Loosanoff and Tommers (1947)[11]
	Increased larval mortality	7.0	Calabrese and Davis (1966)[12]
	Inhibited larval development	6.75	Calabrese and Davis (1966)[12]
	Reduced larval growth	6.75	Calabrese and Davis (1966)[12]
Mercenaria mercenaria	Increased larval mortality	6.5	Calabrese and Davis (1966)[12]
	Inhibited larval development	7.0	Calabrese and Davis (1966)[12]
	Reduced larval growth	6.75	Calabrese and Davis (1966)[12]

Chloride and sulphate concentrations are not changed appreciably (worst case, +0.1% and +3.4%, respectively), against a background of very high concentrations naturally present in seawater. Any increase in phosphorous/phosphates is similarly negligible, while a nitrate increase of 0.2% over baseline is likely.

The suspended particulate concentration may increase by about 1 mg/l, which is unlikely to be appreciably deleterious to marine life.

Trace metals and other contaminants
Sea water contains a very wide array of trace metals and other substances, from both natural and anthropogenic sources. As with excess acidity, it is reasonable to assume that some form of equilibrium is reached, through

buffering, chelation, reaction to form insoluble salts and inclusion in sediment layers. Nonetheless, trace metals, with their ability to be concentrated up through the food chain, are a group of contaminants of some concern. Many of these substances are liable to bioaccumulation, with potentially serious consequences for higher trophic levels in the food web (including humans). In addition, raised levels of trace contaminants can lead to growth inhibition and many other sub-lethal effects in marine algae, phyto- and zooplankton and macro-invertebrates, and promote changes in the food web by favouring the growth of species and forms tolerant of raised trace metal levels.

Wellman Lord process

This is a regenerative FGD process that does not create a waste product stream. SO_2 from flue gas is absorbed in a sodium sulphite solution forming sodium bisulphite; other components of flue gas are not absorbed. After lowering the temperature, the bisulphite is converted to the sodium pyrosulfite, which precipitates. On heating, the chemical reactions are reversed, and sodium pyrosulfite is converted to a concentrated stream of SO_2 and sodium sulphite. The SO_2 can be used for sulphuric acid production or to produce sulphur. This process is little used in power plants.

The decline in SO_2 emissions

Dramatic declines in SO_2 emissions have been reported. The EIA reports emissions from US power plants declined 73% between 2003 and 2015, mainly as a result of the switch to natural gas for electricity generation. Similarly, in the UK emissions from industrial combustion sources have fallen by 96% between 1970 and 2014.

Nitrogen Oxides

Pollution from nitrogen oxides (NOx) causes ground-level ozone, or smog, which can burn lung tissue, exacerbate asthma and make people more susceptible to chronic respiratory diseases. A typical 600 MW (capacity factor = 69%) uncontrolled coal plant emits 10,300 tons of NOx per year. The same coal plant with emissions controls, including selective catalytic reduction technology, emits about 3300 tons of NOx per year.

Selective catalytic reduction is a means of converting NOx with the aid of a catalyst into diatomic nitrogen (N_2) and water. A gaseous reductant, typically anhydrous ammonia, aqueous ammonia or urea, is added to a stream of flue or exhaust gas and is adsorbed onto a catalyst. In power stations, the selective catalytic reduction unit is placed between the furnace economizer and the air heater, and the ammonia is injected into the catalyst chamber (see Fig. 7.11). The temperature of operation is critical.

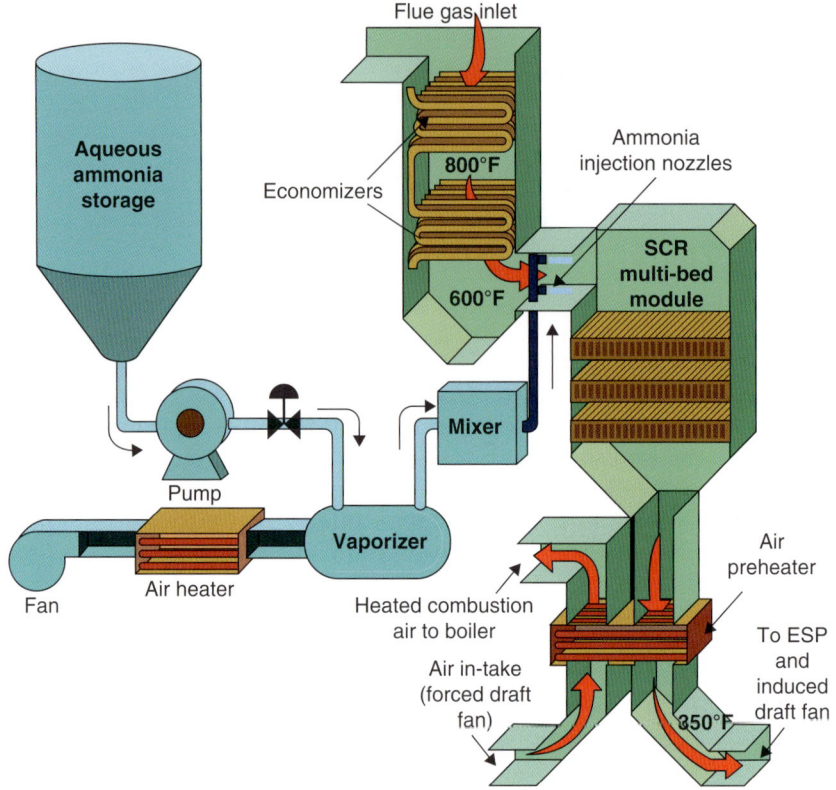

Fig. 7.11. A diagram of a NOx selective catalytic reduction technology fitted to a boiler. (From US Department of Energy, public domain.)

Costs of removal

The costs for selective catalytic reduction equipment to remove nitrogen dioxides are US$124/kW for a 300 MW plant, US$108/kW for a 500 MW plant and US$98/kW for a 700 MW plant. Selective catalytic reduction units are assumed to remove 90% of the NOx.

Mercury

Mercury contamination of aquatic ecosystems and methylmercury bio-accumulation is a global problem. Contamination is generally via atmospheric transport. Using ice-core data, Schuster *et al.* (2002)[13] found that integrated over the past 270-year ice-core history, anthropogenic inputs contributed 52%, volcanic events 6% and background sources 42%. Anthropogenic inputs have increased in importance because over the last 100 years they contributed 70% of the total mercury input. There is some good news in that the last 10 years of this ice-core record showed a decline in atmospheric mercury deposition. Since industrialization began, there has been a gradual build-up of mercury contamination. This has resulted in an increase in mercury levels in fish flesh and the issuing of advisories to limit consumption of fish from many water bodies. For instance, all New Hampshire, USA, surface waters are listed as impaired for mercury because of fish tissue concentrations, which has led to a state-wide fish consumption advisory.

Coal-fired power stations are important sources of mercury. For example, coal plants are responsible for more than half of the US anthropogenic emissions of mercury. Mercury is a trace component of coal and is released with the flue gas. About 60–80% of the mercury present in the flue gas is expected to be caught on the baghouse filters. Thus some 40% to 20% of the mercury in the coal will either pass up the stacks into the atmosphere or, if fitted, enter the flue gas desulphurization system. In modern coal plants, the FGD system can be fitted with alkali/sulphide precipitation systems to precipitate the mercury. This system is probably capable of reliably reducing mercury concentrations to between 1 to 10 µg/l (parts per billion). The Bechtel (2004) study document 24896-000-30L-M21G-00002-000 relating to the Elm Road Generating Station, Wisconsin, indicated that alkali/sulphide precipitation could reduce mercury concentrations below 0.4 ppb; however, these were laboratory tests and should not be viewed as a reliable indication of what could be delivered in practice. The most stringent US water-quality criterion for mercury is the wildlife criterion of 1.3 ng/l (1 µg/l = 1000 ng/l). Thus, even if the alkali/sulphide precipitation system can reduce mercury in the water to about 1 µg/l, it is still well above the most stringent wildlife criterion. Additional technologies can be applied to further remove mercury from the effluent stream, but the Bechtel (2004) report dismissed these as of little proven use, impractical or too expensive.

In the case of the study for the Elm Road Generating Station, Wisconsin, the final conclusion reached by the Wisconsin Electric Power Company in their November 2004 report entitled 'Costs and cost-effectiveness – polishing treatment technologies for mercury removal at ERGS' is that none of the methods that can reduce the discharge below 1 µg/l is cost-effective. They therefore needed a discharge permit of the order of 4 to 1 µg/l (ppb) and a mixing zone. Such discharge levels had been permitted in other localities but may no longer be considered acceptable for the Great Lakes of Canada and the USA.

Notes

[1] Environmental Protection Agency (2000) Notice of Regulatory Determination on Wastes From the Combustion of Fossil Fuels. *Federal Register* 65(99).
[2] National Research Council (2006) *Managing Coal Combustion Residues in Mines*. The National Academies Press, Washington, DC.

[3] Table 9.4. Average costs of existing flue gas desulfurization units operating in electric power sector, 2006–2016. Available at: https://www.eia.gov/electricity/annual/html/epa_09_04.html (accessed December 2017).

[4] Batten, S.D. and Bamber, R.N. (1996) The effects of acidified seawater on the polychaete *Nereis virens* Sars, 1835. *Marine Pollution Bulletin* 32(3), 283–287.

[5] Bamber, R.N. (1985) *The Effects of Acidified Sea-water on Marine Organisms: Experimental Apparatus.* CERL/CEGB memorandum, TPRD/BY/083/M84.

[6] Bamber, R.N. (1987) The effects of acidic sea water on young carpet-shell clams *Venerupis decussata* (L.) (Mollusca: Veneracea). *Journal of Experimental Marine Biology and Ecology* 108(3), 241–260.

[7] Bamber, R.N. (1990) The effects of acidic seawater on three species of lamellibranch mollusc. *Journal of Experimental Marine Biology and Ecology* 143(3), 181–191.

[8] Langford, T.E. and Bamber, R.N. (1984) *The Effects of Acidified Sea Water on Fish.* CERL/CEGB, Technology Planning and Research Division, TPRD/L/2671/N84.

[9] Akberali, H.B., Earnshaw, M.J. and Marriott, K.R.M. (1985) The action of heavy metals on the gametes of the marine mussel, *Mytilus edulis* (L.)-II. Uptake of copper and zinc and their effect on respiration in the sperm and unfertilized egg. *Marine environmental research* 16, 37–59.

[10] Kuwatani, Y., Nishii, T. and Isogai, F. (1969) Effects of pH of culture water on the growth of the Japanese pearl oyster. *Bulletin of the National Research Laboratory* 35, 342–350.

[11] Loosanoff, V.L. and Tommers, F.D. (1947) Effect of low pH upon rate of water pumping of oysters, *Ostrea virginica. The Anatomical Record* 99, 668.

[12] Calabrese, A. and Davis, H.C. (1966) The pH tolerance of embryos and larvae of *Mercenaria mercenaria* and *Crassostrea virginica. The Biological Bulletin* 131, 427–436.

[13] Schuster, P.F., Krabbenhoft, D.P., Naftz, D.L., Cecil, L.D., Olson, M.L., Dewild, J.F., Susong, D.D., Green, J.R. and Abbott, M.L. (2002) Atmospheric mercury deposition during the last 270 years: a glacial ice core record of natural and anthropogenic sources. *Environmental Science & Technology* 36(11), 2303–2310.

8 Gas-Fired Power Plants

A combined-cycle gas-fired power plant is one of the most efficient ways to use fossil fuel for electricity generation. These plants use a gas turbine in conjunction with a heat recovery steam generator and so combine the Brayton cycle of the gas turbine with the Rankine cycle of the steam turbine (see Fig. 8.2). The thermal efficiency of these plants can reach about 60% compared with around 35% for a traditional coal-fired station. The turbines are fuelled with natural gas, syngas or fuel oil.

Gas-fired power stations are quick to build and can often be placed on industrial sites on the outskirts of towns near to where the power is consumed. Open-cycle gas turbine plants, without a steam cycle, are also occasionally built to meet peak demands. They are less efficient and therefore not favoured for base-load use. Gas turbines can be quickly brought up to load and therefore have distinct advantages when combined with intermittent energy sources such as solar and wind farms.

The high energy-conversion efficiency of combined-cycle gas plants results in a waste heat discharge that is smaller than that for comparable nuclear, coal or oil plants. It is therefore more likely that the plant will use closed-cycle cooling (see Chapter 5, p. 43, this volume). However, combined-cycle gas plants may have the same environmental impacts as for steam turbines (see Chapter 5, this volume).

A huge switch to natural gas power generation is now underway. About 80% of all new power generation commissioned in the USA between 2000 and 2010 is gas-fired. This has been stimulated by the decline in the gas price as fracking became popular (see below). In addition, it has the advantage over coal of releasing about half the amount of CO_2 produced by the equivalent coal plant. It therefore has become part of the approach to limit greenhouse gas production.

Fig. 8.1. Grain Power Station, a combined-cycle gas power station in the Thames estuary, England. (Photo courtesy of Richard Seaby, Pisces Conservation Ltd.)

Installed Capacity

In 2014 natural gas-fired power plants generated about 22% of total global electricity output, a figure that is expected to increase further over the next 20 years (see Fig. 1.3, Chapter 1, this volume).

Environmental Issues

The main operational impacts of gas plants are the same as those for other plants using steam turbines (see Chapter 5, this volume). They do have the distinct advantage over coal plants of reduced CO_2 emissions per unit of power generated. For the most efficient combined-cycle plants, emissions are estimated as 365 gCO_2eq/kWh compared with >800 gCO_2eq/kWh for coal plants (see Houses of Parliament POSTnote 383).[1]

The key area of environmental concern unique to gas-fired plants is the impact of natural gas production and fracking in particular.

Environmental Issues Linked to Fracking

This brief summary was first distributed in the Solent Protection Society newsletter (see http://www.solentprotection.org/wp-content/uploads/SPS-Newsletter-Autumn-2016.pdf).

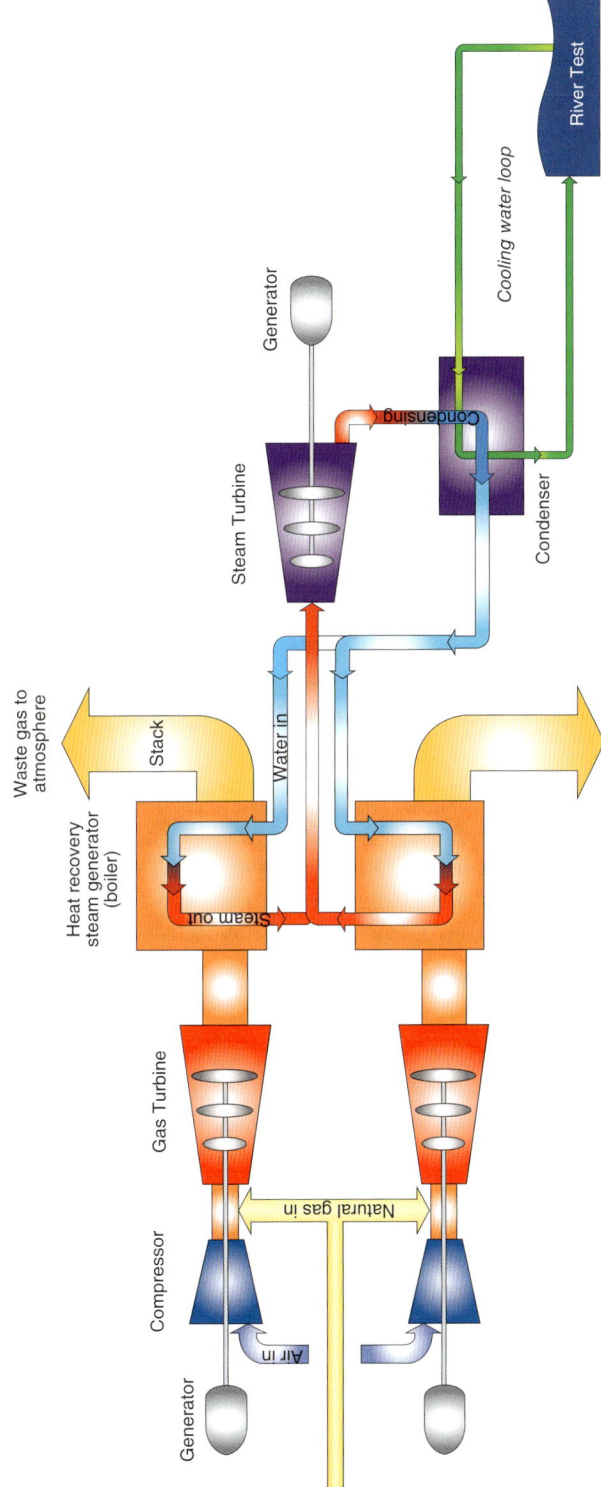

Fig. 8.2. Schematic of the combined-cycle gas turbine process operated at Marchwood Power Station, UK. (Drawing courtesy of Marchwood Power Station.)

What is fracking?

Shale rock formations contain vast quantities of natural gas (which is mostly methane). Until recently, most of this gas was not economically obtainable, because it is held in shale which is far less permeable than the rock formations exploited for conventional gas. This tight gas and oil could not be released in large quantities by simple drilling and pumping from the reservoir. Over the last 20 years two technological innovations have combined to allow extraction of shale gas: high-volume, slick-water hydraulic fracturing (usually shortened to fracking), in which high-pressure water with additives is used to increase fissures in the rock; and precision drilling of wells that can follow the contours of a shale layer closely for 3 km or more at depths of more than 2 km. Drillers first experimented with these two technologies in Texas about 20 years ago. Significant shale gas production in other US states, including Arkansas, Pennsylvania and Louisiana, began in 2007–2009. Outside of North America few shale gas wells have been drilled, but this is set to change.

A brief history

While the individual engineering procedures undertaken during fracking have been long used, fracking of shales for large-scale gas production is a recent development. In the 1940s the hydraulic fracturing process, also known as hydrafrac, began to be developed and, after considerable experimentation, a patent was issued by Halliburton Oil Well Cementing Company (HOWCO) and the first two commercial fracturing treatments were produced. In the early 2000s the modern fracking technique, called horizontal slick water fracking, made the extraction of shale gas economical. This process was first applied in the Texas Barnett Shale (Fig. 8.3).

It is important to note that gas exploration techniques using directional drilling and hydraulic fracturing are not new and have been used across the oil and gas industry for many decades. It would appear that gas and oil resources held in shales that can be fracked are huge.

A summary of the fracking procedure

All wells are different, but a summary of the main steps in well formation (see Fig. 8.4) are as follows.

1. A wellbore is drilled using a drill pipe and bit.
2. Mud is pumped down to the drill to cool and lubricate the drill pipe and bit. The mud helps stabilize the wellbore and helps carry rock fragments to the surface. This is normal oil drilling practice.

Fig. 8.3. Texas Barnett Shale gas drilling rig near Alvarado, Texas. (Photo courtesy of Loadmaster (David R. Tribble), published under a CC BY-SA 3.0 license via Wikimedia Commons.)

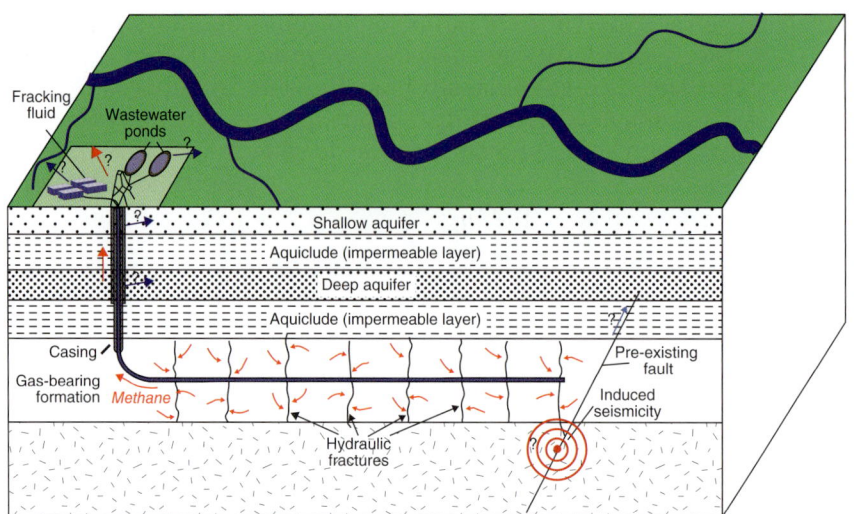

Fig. 8.4. Diagrammatic representation of hydraulic fracturing or fracking. (From Mike Norton, published under a CC BY-SA 3.0 license via Wikimedia Commons.)

3. Drilling passes well below groundwater levels. Typically, thousands of feet of rock separate shale reserves from the lowest groundwater reservoir.

4. The drill pipe and bit are removed and a steel tube called surface casing is set inside the well. The tube stabilizes the sides of the well creating a protective barrier between the well stream and any underground fresh water reservoirs.

5. Cement is pumped into the well through and out of the casing, displacing remaining drilling fluids and securing the casing in place permanently. This cement fills the space between the casing and wellbore and provides a protective seal, keeping outside materials (e.g. ground water) from entering the well flow.

6. The casing is pressure tested to make sure hydrocarbons and other fluids do not seep out into ground water and shallow deposits as they are brought to the surface.

7. Pipe and drill bit are lowered back into the well and drilling continues and another layer of casing and cement are set in place to create further protective barriers. Multiple layers of casing and cement are critical for safe well construction and to protect drinking water.

8. About 500 feet above the hydrocarbon shale formation, a drilling motor with measuring instruments begins drilling at an angle to create a horizontal tunnel running along the targeted layer of gas- or oil-bearing shale. The casing and cementing process continues through the entire length of the now horizontal bore hole.

9. A perforating tool is inserted into the well, creating holes in shale layers, allowing hydrocarbons to enter the well stream.

10. The perforating tool is removed and a fracturing fluid made up of water and sand together with a cocktail of chemicals is pumped at high pressure into the well, opening up tiny fractures deep into the shale. The sand lodges in the cracks keeping them open when the fluid is removed.

11. After the fracturing process is completed, plugs are removed to allow gas or oil to flow to the surface.

To improve the efficiency of the hydraulic fracturing process, chemicals are often added to the water to create the hydraulic fracturing fluid. These chemicals have a number of roles:

- Proppants keep induced fractures open.
- Biocides prevent bacterial growth, which can affect the gas well's productivity.
- Surfactants reduce surface tension, aiding fluid recovery.
- Polymers reduce friction between the hydraulic fracturing fluid.
- Corrosion inhibitors.

The composition of hydraulic fracturing fluids depends on a number of factors, such as the underlying geology, the type of rock to be hydraulically fractured and operational considerations.

Environmental problems arising from fracking

Considerable concerns have been expressed about the environmental damage that fracking can cause. At present these arguments are based on events in the USA. There is, as yet, no first-hand experience of the environmental issues caused by fracking in the UK, for example. This is an important point to remember when reading the claims of both proponents and opponents.

As usual, we can divide impacts into construction and operational phases.

Construction phase
Well construction will require heavy equipment to be moved onto the site. There will therefore be local disturbance to wildlife and residents linked to road construction and the movement and operation of vehicles and machinery. Construction phase noise and air pollution would last for about 100–150 days. The types of inconvenience and pollution caused would be familiar to us all as they differ little in scale or impact from that caused by the construction of large buildings and roads.

Once the well is completed, the heavy machinery is removed. When production commences the site generates little noise or obvious pollution and there is little above-ground structure, as shown in Fig. 8.5.

Fig. 8.5. Well head after all the fracking equipment has been taken off location. (Photo courtesy of Joshua Doubek, published under a CC BY-SA 3.0 license via Wikimedia Commons.)

Operational phase

This is by far the most significant phase in terms of environmental impacts and also lasts for far longer than the 70–150 days of drilling and wellhead construction (maybe as long as 50 years for some wells). The actual physical presence of the well head is often insignificant (Fig. 8.5). However, there are a number of potentially serious environmental issues linked to the release of methane and contamination of water.[2]

In response to emerging public concern about hydraulic fracturing for shale gas extraction, and requests for advice from national and local agencies, the Public Health England (PHE) Centre for Radiation, Chemical and Environmental Hazards reviewed the potential public health impact of direct emissions of chemicals and radioactive material from the extraction of shale gas.[3] PHE considered the following key risks:

- Air pollution, including from stationary on-site sources, and radon.
- Water pollution, including from hydraulic fracturing fluids, flowback water, presence of natural substances, e.g. heavy metals and NORM, and the risk to watercourses or aquifers.
- Land/waste issues, including disposal and treatment of wastewater, muds, etc.

AIR POLLUTION. A review of the peer-reviewed scientific literature and discussions with key agencies did not identify any UK data (published or unpublished) on emissions to air associated with hydraulic fracturing or shale gas extraction. There is one site that has been operating in the UK since 1996 at Elswick, Lancashire, with vertical hydraulic fracturing to release gas in sandstone formations, but no data on air emissions associated with this site appear to have been published. Similarly, no air quality data have been obtained for more recent exploratory drilling for shale gas at Preese Hall, Lancashire.

Information from operations in the USA and Canada indicate that emissions from individual shale gas wells are relatively small, intermittent and not unique to shale gas extraction and related activities.

Radon release does not seem to be especially associated with fracking. Typical levels of radon in natural gas have been reviewed. For typical domestic rates of gas usage with an average UK radon level of about 200 Bq/m^3, the estimated individual annual dose from the use of natural gas for UK residents is estimated at 4 μSv (microsieverts), which is extremely small.

WATER POLLUTION. Disposal of flowback wastewater during shale gas extraction processes presents potential public health risks due to the chemical constituents and large volumes of water involved. In the USA, management of flowback water involves on- or off-site treatment, recycling for reuse or storage. Some operators in the USA have stored flowback water on-site in man-made ponds or open pits. Stored water may be a source of surface water contamination or air pollution due to evaporation of volatile organic compounds. Additionally, evaporation can increase the concentration of dissolved substances and so further increase the potential public health risks associated with storage on-site. The UK regulatory approach will not permit this practice.

Storage of hydraulic fracturing fluids and flowback water on-site may lead to accidents and potential releases to the environment. There is the risk of stored hydraulic fracturing fluids and flowback water entering nearby surface water bodies or infiltrating into the soil and near-surface groundwater as a result of mismanagement. Such leakage may potentially affect drinking water resources if formal accident and contingency plans are not in place (US EPA, 2012).[4]

The potential impacts of shale gas extraction and related activities on UK drinking water sources have been considered in several reviews, most notably by the House of Commons Energy and Climate Change Committee,[3,5] the Tyndall Centre for Climate Change Research (Broderick et al., 2011)[6] and The Royal Society and The Royal Academy of Engineering (The Royal Society, 2012).[7] These reviews noted that groundwater and surface water pollution had been reported in the USA and also recommended appropriate regulatory control to prevent such cases in the UK. The need to demonstrate good well integrity was considered a key regulatory measure.

PHE[8] concluded that 'the potential health impact from single wells is likely to be very small, but the cumulative impacts of many wells in various phases of development in relatively small areas are potentially greater and will need careful scrutiny, during the planning process.'

In addition to public health issues, there are a range of more general environmental and social concerns, the main ones being:

- Occupational health.
- Water usage and water sustainability.
- Energy policy and security.
- Nuisance issues, including noise and odours.
- Seismicity. Small earth tremors caused the shutdown of the first UK fracking operation.
- Wider impacts of shale gas extraction on local employment and the local economy.
- Detailed consideration of the longer term impact of shale gas extraction on climate change. Methane is a greenhouse gas as is the CO_2 produced by methane combustion.
- Community concerns linked to socioeconomic impacts, visual amenity and noise.
- Longer term impact of shale gas extraction on greenhouse gas emissions and their impact on climate change.

Notes

[1] Houses of Parliament (2011) *Carbon Footprint of Electricity Generation.* POSTnote update 383. Available at: http://www.parliament.uk/mps-lords-and-offices/offices/bicameral/post/publications/postnotes/ (accessed December 2017).
[2] Howarth, R.W., Ingraffea, A. and Engelder, T. (2011) Natural gas: should fracking stop? *Nature* 477(7364), 271–275.
[3] GOV.UK (2014) *Shale Gas Extraction: Review of the Potential Public Health Impacts of Exposures to Chemical and Radioactive Pollutants.* Available at: https://www.gov.uk/government/publications/shale-gas-extraction-review-of-the-potential-public-health-impacts-of-exposures-to-chemical-and-radioactive-pollutants (accessed September 2017).
[4] US EPA (2012) Hydraulic Fracturing for Oil and Gas: Impacts from the Hydraulic Fracturing Water Cycle on Drinking Water Resources in the United States (Final Report). U.S. Environmental Protection Agency, Washington, DC, USA. EPA/600/R-16/236F, 2016.
[5] White, E., Fell, M., Smith, L. and Keep, M. (2014) *Shale Gas and Fracking.* Parliamentary Briefing. The Stationery Office, London.
[6] Broderick, J., Anderson, K., Wood, R., Gilbert, P., Sharmina, M. *et al.* (2011) *Shale Gas: An Updated Assessment of Environmental and Climate Change Impacts.* A report commissioned by The Co-operative and undertaken by researchers at the Tyndall Centre. University of Manchester, Manchester, UK.
[7] The Royal Society (2012) Shale Gas Extraction in the UK: A Review of Hydraulic Fracturing. DES2597. Available at: royalsociety.org/policy/projects/shale-gas-extraction (accessed 5 April 2018).
[8] Public Health England (2014) Review of the Potential Public Health Impacts of Exposures to Chemical and Radioactive Pollutants as a Result of the Shale Gas Extraction Process. PHE publications gateway number: 2014007. Available at: https://www.gov.uk/government/uploads/system/uploads/attachment_data/file/332837/PHE-CRCE-009_3-7-14.pdf (accessed 5 March 2018).

9

Wind Turbines and the Effects of Offshore Piling

Wind turbines, which convert wind kinetic energy into electrical power, are a familiar site in many parts of Europe and North America. Wind was a primary source of power until the industrial revolution and windmills were a feature of rural views in many countries. English speakers still tend to refer to electricity-generating wind turbines as windmills. There are a bewildering variety of windmill and wind turbine designs, but horizontal three-bladed designs are by far the commonest design in large-scale wind farms (see Fig. 9.1). It is the ecological impacts of this common design that is the main focus of this chapter. The reason for the popularity of the tri-blade wind turbine machine is because in most conditions they are the most efficient design. Their aerodynamic blades generate lift to drive the blade faster. This is a significant advantage over windmills of either horizontal or vertical-axis design.

Included in this chapter are reviews of ecological impacts of piling and dredging on marine life. These activities often occur during the construction of offshore wind farms; however, they also commonly occur during the construction of other types of coastal power plant.

Growth in Installed Wind Turbines

Worldwide installed wind capacity has grown exponentially since 1997 (Fig. 9.2). In 2016, total global installed capacity was nearly 487 GW, with the fastest growth in new capacity occurring in China, the USA, Germany and India (Table 9.1). While growth may deaccelerate, it is certain to continue for many years to come. The ecological effects of wind turbine operation and installation are therefore likely to become issues of increasing interest.

Fig. 9.1. Newly constructed wind turbines D4 (nearest) to D1 on the Thornton Bank, 28 km offshore, in the Belgian part of the North Sea. The windmills are 157 m (+TAW) high, 184 m above the sea bottom. (© Hans Hillewaert, via Wikimedia Commons.)

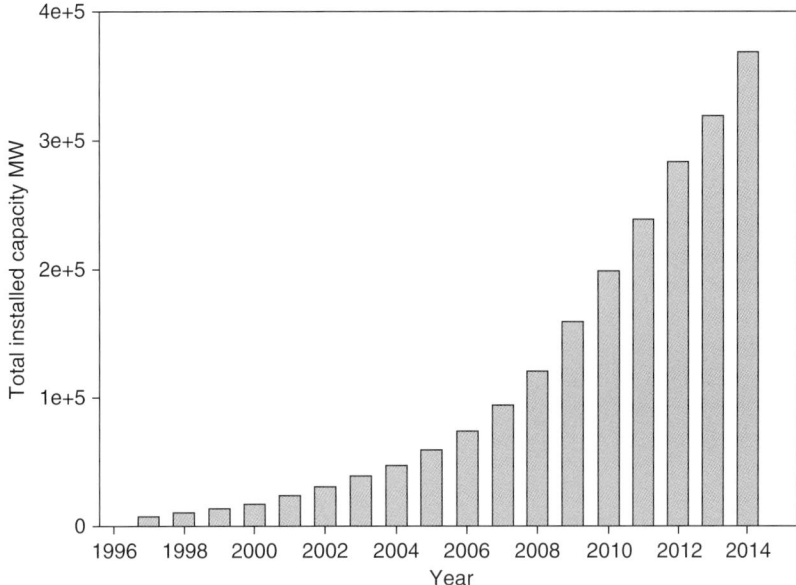

Fig. 9.2. The growth in worldwide wind turbine electricity-generating capacity. (Data from http://www.gwec.net/wp-content/uploads/2015/02/GWEC_GlobalWindStats2014_FINAL_10.2.2015.pdf.)

Atmospheric and Carbon Emissions

While there are no significant atmospheric emissions associated with operating wind turbines, there are emissions during materials production, materials transportation, on-site construction and assembly, operation and maintenance, and decommissioning. Estimates of total global warming emissions depend on a number of factors, including wind speed, per cent of time the wind is blowing and the material composition of the wind turbine. Most estimates of wind turbine life-cycle global warming emissions are between 0.02 and 0.04 pounds of carbon dioxide equivalent per kilowatt-hour. Life-cycle global warming emissions for natural-gas-generated electricity are between 0.6 and 2 pounds of carbon dioxide equivalent per kilowatt-hour and for coal-generated electricity 1.4 to 3.6 pounds of carbon dioxide equivalent per kilowatt-hour.[1]

Ecological Impacts of Large-Scale Offshore Wind Farms

Ecological impacts on marine wildlife relating to large-scale offshore wind generation are discussed by Bergstrom *et al.* (2014).[2] The areas of concern identified are discussed below. The bulk of our present knowledge has been acquired over the last 10 years.

Table 9.1. The installed wind turbine capacity in MW for the years 2001–2016 for the 20 countries with the greatest installed capacity in 2016. (Data from http://www.gwec.net/.)

		Installed wind turbine capacity (MW)					
		2011	2012	2013	2014	2015	2016
1	China	62,733	75,564	91,412	114,763	145,104	168,690
2	USA	46,919	60,007	61,110	65,879	74,472	82,183
3	Germany	29,060	31,332	34,250	39,165	44,947	50,019
4	India	16,084	18,421	20,150	22,465	27,151	28,665
5	Spain	21,674	22,796	22,959	22,987	23,025	23,075
6	United Kingdom	6,540	8,445	10,711	12,440	13,603	14,542
7	France	6,800	7,196	8,243	9,285	10,358	12,065
8	Canada	5,265	6,200	7,823	9,694	11,205	11,898
9	Brazil	1,509	2,508	3,466	5,939	8,715	10,740
10	Italy	6,747	8,144	8,558	8,663	8,958	9,257
11	Sweden	2,970	3,745	4,382	5,425	6,025	6,519
12	Turkey	1,799	2,312	2,958	3,763	4,718	6,081
13	Poland	1,616	2,497	3,390	3,834	5,100	5,782
14	Portugal	4,083	4,525	4,730	4,914	5,079	5,316
15	Denmark	3,871	4,162	4,807	4,845	5,063	5,227
16	Netherlands	2,328	2,391	2,671	2,805	3,431	4,328
17	Australia	2,176	2,584	3,239	3,806	4,187	4,327
18	Mexico	873	1,370	1,859	2,551	3,073	3,527
19	Japan	2,501	2,614	2,669	2,789	3,038	3,234
20	Romania	982	1,905	2,599	2,953	2,976	3,028
	World total capacity	238,035	282,482	318,596	369,553	432,419	

Construction phase impacts

The key issues in many wind farm constructions concern acoustic disturbances and increased sediment dispersal linked to engineering works required to anchor the turbines to the seabed.

Piling

Pile driving is used on offshore wind farms based on monopiles or jacketed foundations. While piling is a feature of many construction activities in coastal waters, including those linked to the building of coal, oil and nuclear plants, the subject is covered in depth in this chapter because it is a key construction phase impact of many offshore wind farms. Pile driving may generate appreciable levels of underwater noise. Noise levels close to the pile driver (within about 5 m) can exceed the levels that will hurt or kill fish[3] (peak quoted values 218 dB) and other marine life. Given this risk, the level of noise generated by impact and vibratory piling and the impact on fish and marine mammals is considered in detail at the end of this chapter.

Construction phase impacts with gravity foundations

Unlike percussive piling, the installation of gravity foundations does not generate high sound levels. However, there is an appreciable level of disturbance caused by boat noise and dredging. Dredging is a locally destructive activity that can have a range of ecological impacts that are considered in more detail below. As with the consideration of piling, dredging is a widely carried out procedure not only associated with offshore wind farm construction. Major dredging campaigns can occur during harbour construction and for the construction of cooling water intakes and outfalls at conventional and nuclear power plants.

Dredging can cause a variety of impacts on fish and other marine life, arising from sediment plumes, oxygen sags, increased contaminant levels, entrainment mortality and damage to important habitat. These are considered in turn in this section.

THE EFFECTS OF SUSPENDED SEDIMENT.   The effect of suspended sediments on fish has been found to be the product of the suspended solid concentration and the duration of the exposure. Because the behavioural responses of many fish to elevated suspended sediment concentration (SSC) levels are unknown, it is not possible to predict with certainty the time of exposure. Inshore fish live in a habitat with changing amounts of suspended sediment. It varies daily and seasonally with tidal flows, currents, river inputs and storms. It also varies considerably between estuaries. Exceedingly turbid estuaries such as the River Severn and Bristol Channel, UK, where suspended solids can reach 3000 mg/l, still hold rich fish communities. However, some species do not seem to be able to withstand high SSC levels and this could explain the lack of smelt (*Osmerus eperlangius*) in the Severn Estuary. Fish are assumed to have behavioural and physiological mechanisms to deal with living in turbid waters. Dredging produces suspended sediment plumes that can differ in timing, scope intensity and duration from that found naturally. These may be outside the normal tolerances for the fish living within or passing through the area affected.

Most adult or juvenile fish are sufficiently mobile to leave waters with intolerably high SSC and will therefore likely only be exposed for short periods. Some benthic adult fish and most egg and larval stages have limited mobility and may be exposed to the high SSC for extended periods. Marine fish either lay pelagic buoyant eggs moved passively by the water currents, or demersal eggs, which are buried in or attached to the seabed. The lack of spawning within estuaries compared with adjacent fresh or coastal waters is probably linked to the vulnerability of the eggs and larvae to high SSC and subsequent smothering following settlement.

Fish obtain oxygen by passing water over gills, which are easily damaged. High SSC can block gills, so that in extreme cases the fish suffocates. The levels of SSC predicted to be produced in the proposed dredging operation are not high enough to block the gills of fish completely, so suffocation should not occur. A possible minor cost to the fish caused by elevated SSC is an increase in mucus production on the gills and an increase in gill clearing. Both of these have a metabolic cost and might, if high SSC levels lasted for extended periods, affect the energy budget. In laboratory tests involving Pacific salmon and juvenile trout, fish lived for 3–4 weeks in SSC of 300–750 mg/l, increased to 2300–6500 mg/l for short periods. Sub-lethal pathological effects included increased mucus production over the body and gills plus evidence of actual gill damage at greatly elevated SSC (FARL, 1995).[4]

Work in America has clearly demonstrated that the eggs and larvae of estuarine fish are sensitive to high SSC (Clarke *et al.* 2000).[5] The herring (*Clupea harengus*) is a bottom-spawning fish and Kiorboe *et al.* (1981)[6] found that its egg development was not impaired after 1-day exposures to SSC levels of 300 and 500 mg/l. The potential vulnerability to smothering of benthic eggs was demonstrated by Messieh *et al.* (1981),[7] who showed that the burial of Atlantic herring eggs under even a thin veneer of sediment caused substantial mortality.

Some fish are visual predators and increased SSC levels may reduce their ability to feed in affected areas.

THE REDUCTION OF DISSOLVED OXYGEN BY SUSPENDED SEDIMENTS. If the dredged sediments have a high biological or chemical oxygen demand, then it is possible that the oxygen concentration of the water near the dredging site will be reduced. Most fish would respond by avoiding areas with low oxygen, with no long-lasting effect.

RELEASE OF CONTAMINANTS. The sediments in estuaries are often contaminated with chemicals released from industrial processes, water run-off and sewage waste. Many chemicals have an affinity for the fine particles found in marine sediment and if present can potentially be remobilized by dredging, when they may enter the estuarine food chain.

HABITAT DAMAGE. Dredging disturbs and modifies local habitats, which may take an extended period to recover.

ENTRAINMENT OF FISH BY THE DREDGE. Entrainment is the removal of animals with the water and sediment during the operation of the dredging machinery. Both demersal and pelagic fish eggs and larvae are susceptible to entrainment in suction dredges as they are unable to escape from the flow in the vicinity of the suction pipe (McNair and Banks 1986).[8] Entrainment will result in the almost certain death for fish by burial in the sediment, pressure changes or abrasion.

The capture of juvenile and adult fish will be much lower because they are stronger swimmers and thus able to escape. Avoidance was demonstrated by Armstrong *et al.* (1982),[9] who reported species-specific rates ranging from 0.001 to 0.135 fish/cubic yard. While fish up to 234 mm in length were entrained, trawl data indicated that many species were apparently capable of escaping the dredge. Fish up to 240 mm have been reported to be impinged, but fish more than 50 mm in length are unlikely to be caught in large numbers.

The few available estimates for entrainment rates of fish during marine dredging available are for North American species and are summarized in Table 9.2.

Table 9.2. Estimated entrainment rates of estuarine fish during dredging expressed in numbers of fish per cubic yard of sediment. (From Reine and Clark, 1998[10]; adapted from Larson and Moehl, 1990 and McGraw and Armstrong, 1990.)

Species	Hopper 1 (fish/cubic yard)	Hopper 2 (fish/cubic yard)	Pipeline 2 (fish/cubic yard)
Anchovy (Engraulididae)	0.008	0.001	–
Northern anchovy (*Engraulis mordax*)	–	0.018	–
Herring (Clupeiformes)	0.008	–	–
Arrowtooth flounder (*Atheresthes stomias*)	–	0.008–0.022	–
Starry flounder (*Platichthys stellatus*)	–	0.001–0.002	–
English sole (*Pleuronectes vetulus*)	–	0.006–0.035	0.001–0.003
Sand sole (*Psettichthys melanostictus*)	–	0.001–0.016	–
Slender sole (*Lyopsetta exilis*)	–	0.001	–
Pacific sanddab (*Citharichthys sordidus*)	–	0.004–0.076	–
Speckled sanddab (*Citharichthys stigmaeus*)	–	0.003	–
Flatfish (Pleuronectiformes)	0.008	0.001–0.028	–
Buffalo sculpin (*Enophrys bison*)	–	0.006	–
Prickly sculpin (*Cottus asper*)	–	0.020	0.004
Pacific staghorn sculpin (*Leptocottus armatus*)	0.003	0.007–0.092	0.001–0.037
Cabezon (*Scorpaenichthys marmoratus*)	<0.001	–	–
Kelp greenling (*Hexagrammos decagrammus*)	–	0.001	–
Lingcod (*Ophiodon elongatus*)	–	0.001–0.002	–
Poacher (Agonidae)	–0.009	–	–
Warty poacher (*Occella verrucosa*)	–	0.009	–
Snailfish (Cyclopteridae)	–	0.001	–
Showy snailfish (*Liparis pulchellus*)	0.002	–	–
Pacific sandfish (*Trichodon trichodon*)	<0.001	0.002	–
Pacific sand lance (*Ammodytes hexapterus*)	0.341	0.036–0.594	–
Saddleback gunnel (*Pholis ornata*)	–	0.001–0.005	0.023
Snake prickleback (*Lumpenus sagitta*)	–	0.003–0.135	–
Surfperch (Embiotocidae)	<0.001	0.001	–
Eulachon (*Thaleichthys pacificus*)	0.002	–	–
Chum salmon (*Oncorhynchus keta*)	–	–	0.008
Smelt (Osmeridae)	–	0.009	–
Pipefish (Syngnathidae)	–	0.008	–
Bay pipefish (*Syngathus leptorhynchus*)	–	0.006	–
Three-spined stickleback (*Gasterosteus aculeatus*)	–	–	0.004
Big skate (*Raja binoculata*)	<0.001	–	–
Longnose skate (*Raja rhina*)	–	0.003	–
Pacific tomcod (*Microgadus proximus*)	<0.001	0.001–0.008	–
Spiny dogfish (*Squalus acanthias*)	<0.001	–	–

Operational phase impacts of offshore wind farms

Seabed and underwater structure use and exclusion

Offshore wind farms create a hard surface habitat that is in effect an artificial reef. Artificial reefs in areas with little seabed structure can lead to an increase in biodiversity. Increased species abundances have been observed in several studies close to offshore wind farm foundations. In a Baltic Sea study, Wilhelmsson et al. (2006)[11] found demersal fish abundance was highest in the vicinity of the turbines when compared with surrounding areas. However, species richness and alpha diversity were similar. The monopiles of the turbines had a lower species richness than the seabed and was dominated by mussels and barnacles; however, the associated fish community had a greater abundance. They concluded that 'offshore windfarms may function as combined artificial reefs and fish aggregation devices for small demersal fish'. As offshore wind farms offer a novel, artificial habitat, they run the risk of attracting or supporting non-native species. Bulleri and Airoldi (2005)[12] argued that artificial substrates had facilitated the spread of a non-native alga, *Codium fragile*, in the Adriatic.

Reubens et al. (2011)[13] concluded from studies in the North Sea that pout (*Trisopterus luscus*) and cod (*Gadus morhua*) juveniles were present in large numbers in the reefs at the base of wind turbines, where they fed on epifauna. However, they noted that these aggregations had not, as yet, resulted in a detectable increase in pout or cod at a regional level. They did not dismiss the possibility of long-term change and suggested continued monitoring.

Not all seafarers view offshore wind farms positively: other vessels and fishermen can be excluded for safety reasons and, even when permitted to enter, they may no longer be able to trawl in traditional areas. Vessel movements would typically be prohibited from an offshore wind farm site during construction, maintenance and decommissioning. Fisheries exclusion is likely to increase local species abundances. Exclusion of shipping also carries environmental costs in terms of potentially longer travel distances and an increased risk of collision and subsequent pollution.

Lindeboom et al. (2011)[14] summarize what is likely to be the general situation. Wind farms create a new type of habitat that supports a higher biodiversity of benthic organisms than the surrounding, typically soft sediments. This community generates increased use of the area by the benthos, fish, marine mammals and some bird species. However, other species, including some birds, avoid the areas.

Electrical fields

The effects of an electromagnetic field generated by undersea cables is discussed by Gill et al. (2012).[15] They highlight our present lack of knowledge, but suggest that electromagnetic fields from subsea cables may interact with migrating eel *Anguilla* sp., and possibly other diadromous fishes, to temporarily change their swimming direction. Whether this represents a biologically significant effect is unknown.

Noise
Vibrations generated by the gearbox mesh and the generator typically cause underwater noise of 80–150 dB re 1 µPa, at wavelengths that are within hearing range of both fish and mammals. In addition, acoustic disturbance may increase due to increased vessel movements for service and maintenance. There is, at present, no evidence of negative effects linked to wind turbine noise (Bergström *et al.*, 2014).[16]

Impacts on bird, bats and other flying animals
This important area of concern is discussed in more detail in the section on onshore wind farms below, for which we have considerably more useful data. It is almost impossible to detect fatalities at offshore facilities and so we have no data on which to assess the impact.

With respect to birds, the key concerns are collisions, barrier effects and habitat loss. Particular concern is focused on species undertaking regular seasonal migrations. For example, hundreds of millions of birds cross the North and Baltic Seas at least twice every year. A study by Hüppop *et al.* (2006)[17] concluded that almost half of these birds fly at altitudes at which they could be killed by a turbine. They also showed that, especially under poor visibility, terrestrial birds are attracted by illuminated offshore structures and that some species, particularly passerines, collide in large numbers. They argued for: (i) the abandonment of wind farms in zones with dense migration; (ii) turning off turbines on nights predicted to have adverse weather and high migration intensity; and (iii) actions to make wind turbines more recognizable to birds, including modification of the illumination to intermittent rather than continuous light.

While the impacts of offshore wind farms on bats are not understood, it is known that bats do fly offshore and are therefore vulnerable to harm. For example, they regularly migrate across the Baltic and North Sea,[18] where there are extensive wind farms. Migration by bats over water has also been observed in North America. For example, Johnson *et al.* (2011)[19] report the presence of five migrating bat species on a barrier island off the coast of Maryland, USA.

Insect concentration in the vicinity of wind turbines is known to occur at onshore wind farms; the situation offshore is unknown.

Large-Scale Onshore Wind Farms

Onshore wind farms are often unpopular because of damage to the landscape and visual amenity. This is a matter of personal taste rather than ecology and is not discussed further here.

Land use

A key issue is land use. Typically, horizontal wind turbines (see Fig. 9.3) must be spaced 5 to 10 rotor diameters apart. The turbines and associated

Fig. 9.3. An onshore wind farm, Out Newton Wind Farm, UK. (Photo courtesy of Dr Richard Seaby, Pisces Conservation Ltd.)

infrastructure, including roads and transmission lines, therefore only occupy a small portion of the total area of a wind farm. A survey by the
National Renewable Energy Laboratory of large wind facilities in the USA
found that they use between 30 and 141 acres per megawatt of power
output capacity. However, less than 1 acre per megawatt is permanently
disturbed and less than 3.5 acres per megawatt are temporarily disturbed
during construction.[20] The remainder of the land can be used for agriculture and other purposes. Wind turbines are also frequently placed in
commercial and industrial locations such as ports, which reduces land
use concerns.

Wildlife impacts

The key areas of concern relate to operational impacts on flying organisms, particularly birds, bats and insects. Each of these groups is considered in turn.

Birds

Bird deaths and bird habitat loss linked to wind turbines are particularly
contentious issues and bring into focus deep divisions within wildlife
conservation organizations. Perhaps the most striking example of this
conflict occurs in the UK, where the Royal Society for the Protection of
Birds (RSPB) concluded that 'Wind power has a significant role to play in
the UK's fight against climate change. With the right strategic approach,
it can be expanded without detrimental effects on important bird populations.'[21] Note a common feature in discussions of wind farm impacts is
the belief that climate change threats are sufficiently severe to trump the
immediate issues linked to local harm.

It is clear that the risks to birds depends on many factors, including
the design and size of the turbines and their location.

Wind farms can have a variety of adverse effects. The four main impacts identified by Drewitt and Langston (2006)[22] were:

- Disturbance and displacement from desirable habitat (Pearce-Higgins
  et al., 2009).[23]
- Barrier effects – the disruption of favoured flight paths.
- Collision risk (e.g. Hotker et al., 2006).[24]
- Habitat loss or damage.

The number of birds directly killed by collision with turbine blades
is not insignificant. Smallwood (2013)[25] estimated 573,000 bird fatalities/
year at 51,630 megawatt (MW) of installed wind-energy capacity in the
USA in 2012. This number included a worrying 83,000 raptor (birds of
prey) fatalities. In Europe and the USA, it is the loss of large raptors that
has attracted the most concern. The Pine Tree Wind energy project near
Tehachapi, California has been considered to have particularly high raptor
mortality rates and has killed eight golden eagles according to the US Fish

and Wildlife Service. Farfán *et al.* (2017)[26] note that most studies on the effects of wind farms on birds focus on large species and those of conservation concern. They present data on the abundance of birds in the vicinity of a wind farm in an upland habitat in southern Spain both immediately after installation and 6.5 years post-construction. They observed 11 raptors and 38 non-raptor species, of which 30 were passerines. They concluded that while raptor numbers recovered from initial disturbance to levels only slightly lower than those pre-construction, the numbers of non-raptors significantly declined. They noted that while numbers had only slightly decreased, the turbines at this site did act as a barrier to raptor flight. It is also notable, given the significant declines in small bird numbers, that during their study they only observed one bird killed by collision with the blades. It is clear that simply stating that killed birds are rarely observed cannot be used to argue that adverse impacts are not occurring.

Bats

The impacts of wind turbines on bats has been recently reviewed by Arnett *et al.* (2016).[27] They highlight the seriousness of the situation. Bats are killed in a variety of ways, by blunt force trauma, barotrauma and through inner ear damage and other injuries not readily noticed during examination of carcasses in the field. To gain some appreciation of the scale of the problem, the annual mortality rates reported from European and North American studies are summarized in Table 9.3. Arnett *et al.* (2016) noted the 'alarming' lack of data from Mexico, Central and South America, the Caribbean, Africa, New Zealand and Australia. They found no information on bat fatalities at wind farms in mainland Asia; the situation in China is particularly concerning given the rapid growth of wind generation in this country (see Table 9.1).

It is far from clear what impact these levels of mortality are having on bat populations. However, some idea of the losses can be gained from German data. An estimated 10–12 bats are killed annually at each wind turbine in Germany, which suggests that, if all wind turbines are equally destructive, about 200,000 bats are killed annually at onshore wind turbines in Germany alone. These numbers are sufficient to produce concern for

Table 9.3. Wind turbine annual mortality rates for bats from different habitats and geographical regions. (Data from Arnett *et al.*, 2016.[27])

Region	Habitat	Annual death rate per MW installed capacity
USA and Canada	Northeastern deciduous forest	6.1–10.5
USA and Canada	Midwestern deciduous forest/agricultural	4.9–11
USA and Canada	Great Plains	6
USA	Great Basin/Southwest Desert region	1–1.8
Germany	Black Forest	10.5
Europe	Agricultural land	0.6–5.3

future populations because bats are long-lived and have a low fecundity and so cannot quickly replace such losses.

Why bats are vulnerable to wind turbines is unclear. Kunz *et al.* (2007)[28] discuss various hypotheses, as do Cryan and Barclay (2009).[29] There is evidence that collisions are not chance events – bats may be attracted to turbines either as a roost, as a gathering point during the breeding season or to hunt insects concentrated near the blades. Arnett *et al.* (2016) believe that bats that regularly move and feed in more open air-space are most vulnerable; the species most often killed in Europe are aerial-hawking, relatively fast-flying, open-air species.

REDUCING BAT MORTALITIES.   As most bat fatalities in temperate countries occur during relatively low-wind conditions in late summer, restricting turbine operation in light wind conditions can produce an appreciable reduction in bat deaths. One simple approach is to increase the wind speed at which turbines start to operate during periods of the year when bats are particularly vulnerable. Such approaches may reduce mortalities by 50–90%. The use of ultrasonic sound and radars has also been proposed. Changing to other more bat-friendly turbine designs would also be possible but is unlikely to happen. Recently, more complex sets of rules to determine turbine operation based on a number of parameters, including temperature, wind speed, season, time of day and known bats, have been developed. These rules cannot be generalized because they are tailored to the situation at a specific site and turbine design.

If bats are attracted by the availability of insects, then it may be possible to reduce the death rate by painting turbines in colours less attractive to flying insects (see below).

Insects

That wind turbines can kill large numbers of insects is supported by the remarkable fact that insect bodies adhering to the blade leading edges have been implicated in halving turbine power output in high winds (Corten and Veldkamp, 2001).[30] It is well established that insects can be attracted to wind turbines and the degree of attractiveness can be altered by the paint colour of the turbines. It is known that the common turbine colours pure white and light grey both attract insects, as does UV reflecting paint (Long *et al.* 2011).[31] Wind turbines are of sufficient size to interfere with flying insect migrations; for example, monarch butterflies in North America have been reported as killed by wind turbines.

Decommissioning and Recycling of Wind Turbines

It is often assumed that wind energy is a clean, renewable energy source without any adverse emissions. However, this does not fully consider the construction and decommissioning phases. Liu and Barlow (2017)[32] point out that the blades, one of the most important components in the wind turbines, are now made of unrecyclable composite. They estimate the scale of the future problem as 43 million tonnes of blade waste worldwide by 2050, with China possessing 40% of the waste, Europe 25%, the USA 16% and the rest of the world 19%.

Small-Scale Wind Generation

There are many small-scale wind turbine systems available, most of which appear to have small or even negligible ecological impacts when operating. A typical example of a domestic design is the RidgeBlade Wind Turbine (https://ridgeblade.ca/). Another smaller scale design is the Archimedes windmill (https://www.thearchimedes.com/), which is marketed as a low-noise, bird- and bat-friendly design suitable for urban spaces.

The Ecological Impact of Piling

Studies by Nedwell *et al.* (2003)[33] in Southampton Water and Nedwell and Edwards (2002)[34] at Littlehampton give data on the sound levels and the potential impact on fish and diving mammals.

The sound characteristics of impact and vibratory piling

Impact piling

CASE 1: PILING FOR THE RED FUNNEL SOUTHAMPTON TERMINAL. A total of ten tube piles of 508 and 914 mm diameter were driven for this project. However, as the harbour held soft sediments, percussive piling was only used at the end of the work for the final driving of three piles for dynamic testing purposes. The observed change in water pressure at a distance of 96.3 m from the pile driver is shown in Fig. 9.4.

For this study, the authors also computed the pressure history weighted for the sensitivity of salmon (Fig. 9.5).

The variation in peak-to-peak pressure changed from about 195 dB (relative to 1 μP) at the pile driver, to about 152 dB at a distance of about 240 m (Fig. 9.6).

In this study Nedwell *et al.* (2003) found a linear decline in sound pressure with distance (measured in metres) described by the equation:

Sound pressure level = Source level − 0.15 × Distance

Thus, a source level of 194 dB had declined to 150 dB (the safe threshold for no physical effects) within 295 m.

CASE 2: PILING AT THE COUNTY WHARF, ARUN RIVER, LITTLEHAMPTON. For this project, a total of 41 tube piles were driven, using both vibratory and impact piling. The temporal change in pressure was the same as that observed in Southampton Water (see above), with an exponential decline in pressure after each strike (Fig. 9.7).

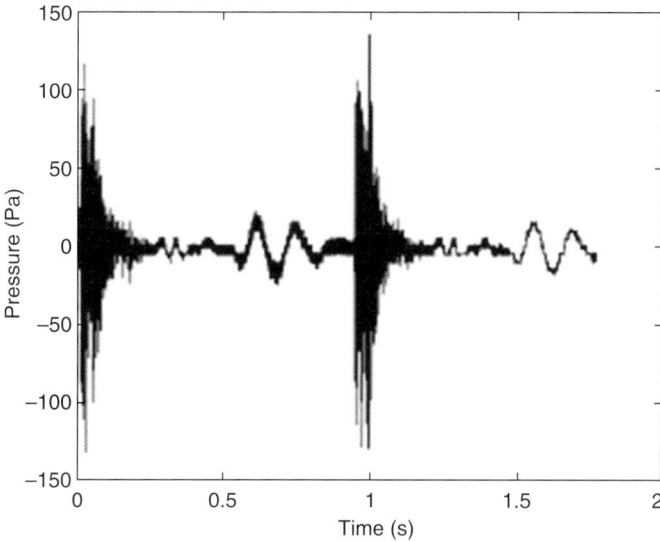

Fig. 9.4. The change in water pressure through time during impact piling at Southampton. (From Nedwell *et al.*, 2003.[33])

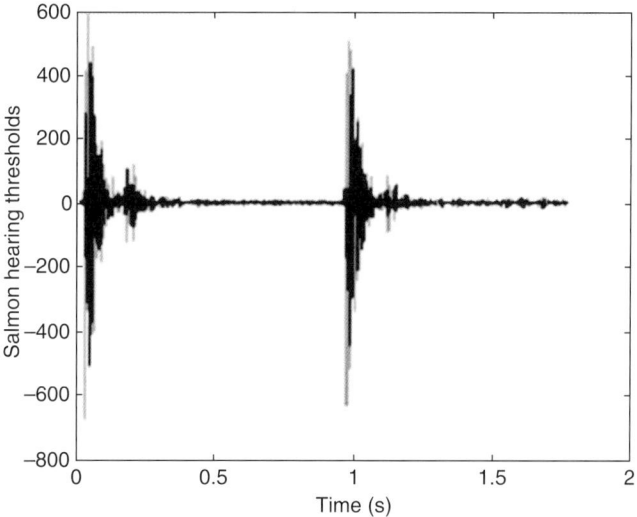

Fig. 9.5. The change in water pressure through time during percussive piling at Southampton, weighted for the hearing sensitivity of salmon. (From Nedwell *et al.*, 2003.[33])

A plot of the change in peak-to-peak pressure with distance from the source is shown in Fig. 9.8.

As in the case of the Southampton Water study, Nedwell and Edwards (2002) considered that a linear equation best described the decline in pressure with distance:

Sound pressure level = Source level − 0.07 × Distance

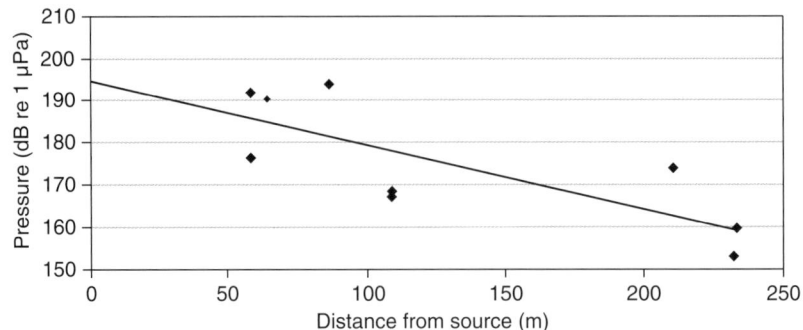

Fig. 9.6. The change in peak-to-peak pressure with distance from the impact piler. (From Nedwell *et al.*, 2003.[33])

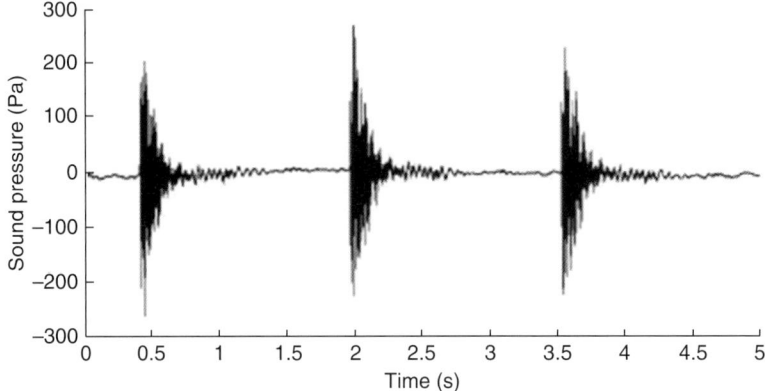

Fig. 9.7. The change in water pressure through time during impact piling at Littlehampton. (From Nedwell and Edwards, 2002.[34])

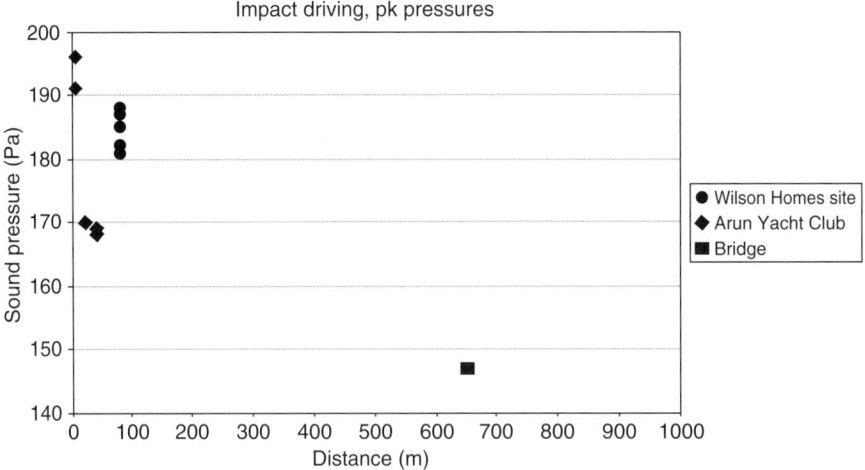

Fig. 9.8. The change in peak-to-peak pressure with distance from the impact piler. (From Nedwell and Edwards, 2002.[34])

NOISE FROM IMPACT PILING IN DEEPER WATERS.   The sound levels reported by Nedwell *et al.* (2003) are lower than the maximum of 246 dB which have been reported for impact piling in deep water. Further, in deep, open waters the decline in pressure with distance is usually modelled as a geometric loss because the decline is primarily the result of dispersion over an ever-increasing area. Within restricted waters and harbours, it would seem the transmission loss is primarily caused by absorption, resulting in a linear decline with distance.

Vibratory piling

CASE 1: PILING FOR THE RED FUNNEL SOUTHAMPTON TERMINAL.   Nedwell *et al.* (2003)[35] recorded the change in sound pressure through time at Town Quay, Southampton approximately 417.4 m from vibratory piling. A total of ten tube piles of 508 and 914 mm diameter were driven for this project, into relatively soft sediments. Percussive piling was only used at the end of the work for the final driving of three piles for dynamic testing purposes. The results showing the total sound pressure variation and the sound pressure frequency weighted for the hearing threshold of salmon are given in Fig. 9.9. The figure clearly demonstrates the considerable variation in sound linked to passing vessels that swamps the variation in sound caused by vibratory piling, which could not be detected in the signal.

The authors concluded that this study clearly demonstrated that salmon cannot detect vibratory piling activity 400 m from the source.

CASE 2: PILING AT THE COUNTY WHARF, ARUN RIVER, LITTLEHAMPTON.   A total of 41 tube piles were driven, using both vibratory and impact piling. The temporal variation

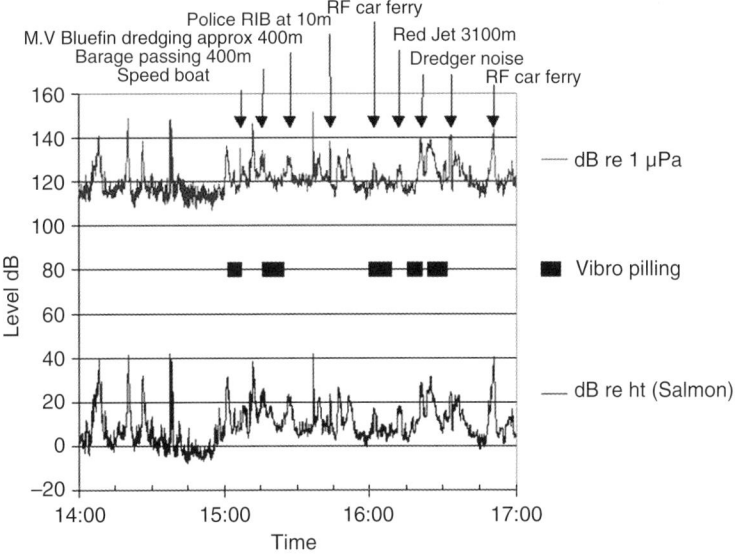

Fig. 9.9. The variation in sound pressure at Town Quay, Southampton, about 417 m from vibratory pile driving. (From Nedwell *et al.*, 2003.[35])

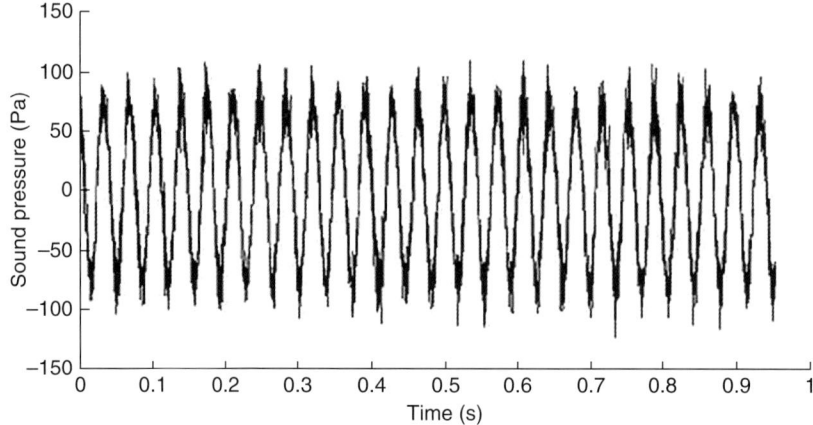

Fig. 9.10. The temporal change in sound pressure during vibratory piling at Littlehampton. (Data from Nedwell and Edwards, 2002.[34])

in sound pressure in the vicinity of the piling is shown in Fig. 9.10 and the frequency spectrum of the sound in Fig. 9.11, in the 0–500 Hz and 0–10 kHz and 0–100 kHz ranges. At a distance of 24 m from the pile, the RMS sound level was 152 dB re 1 µP.

A plot of the change in sound pressure with distance from the source is shown in Fig. 9.12. In this study, it was not possible to determine the transmission loss with distance because of the considerable variation in the sound pressures recorded.

The hearing of fish

All fish appear to be sensitive to sound in the frequency range 50–3000 Hz. This sensitivity is normally presented as an audiogram that plots the sound pressure threshold against the frequency. Examples for a number of fish, including salmon, are shown in Fig. 9.13, which shows the degree of variation between species. In fish, sound is detected by the otolith organs of the inner ears. In addition, the lateral line possesses particle motion detectors that are sensitive to low frequency sound in the range 20–500 Hz. An important determinant of the sensitivity of fish to sound is the presence of a swim bladder and its proximity to the inner ear (Blaxter, 1981).[39]

The response of a fish to sound will, in part, depend on the level of background noise. Work carried out by Buerkle (1968)[37] on the response threshold of cod indicated that a sound must be at least 20 dB above background noise levels before the fish would respond, and at frequencies above 300 Hz it may need to be >40 dB above background. The influence of background noise will clearly be important in any harbour or area with frequent shipping movements.

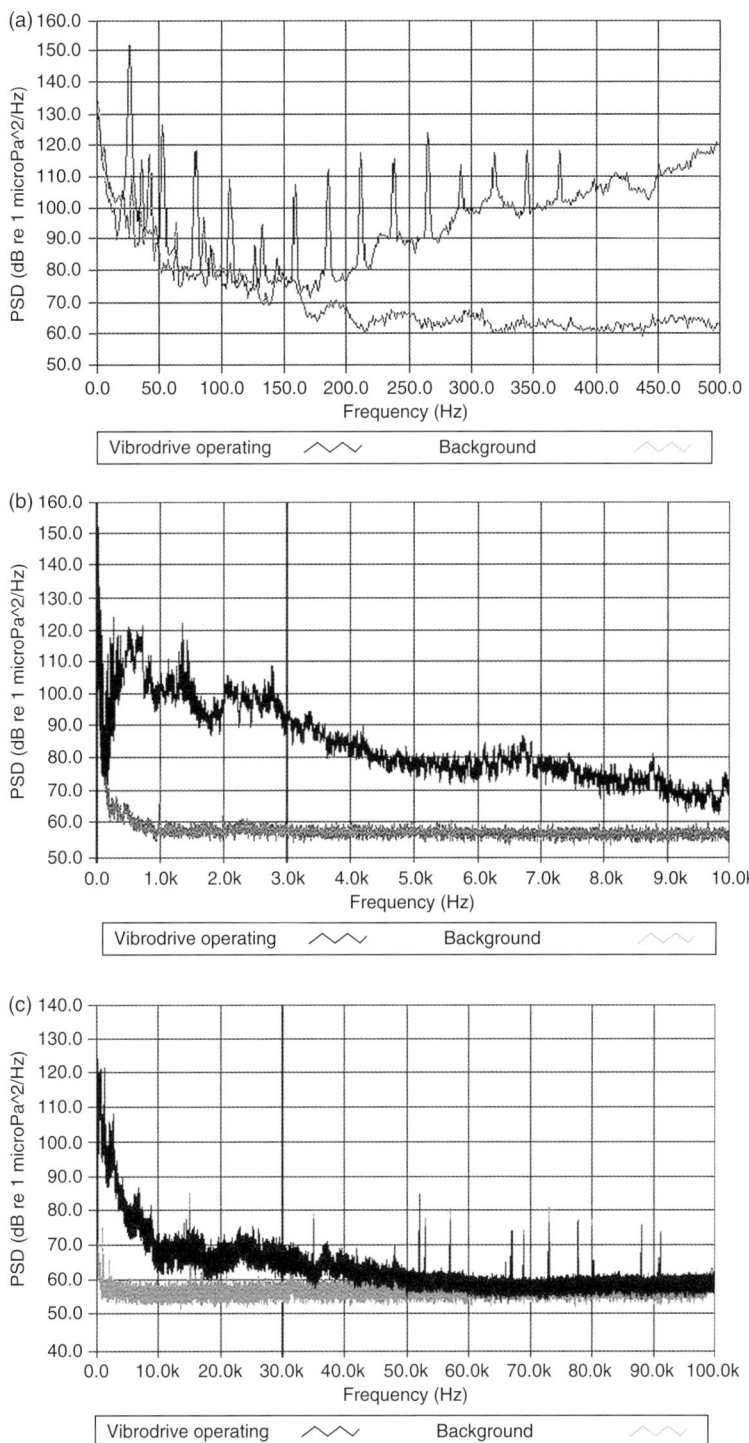

Fig. 9.11. (a–c). The sound spectrum from vibratory piling, Littlehampton, in 0–500 Hz, 0–10 kHz and 0–100 kHz ranges. (Data from Nedwell and Edwards, 2002.[34])

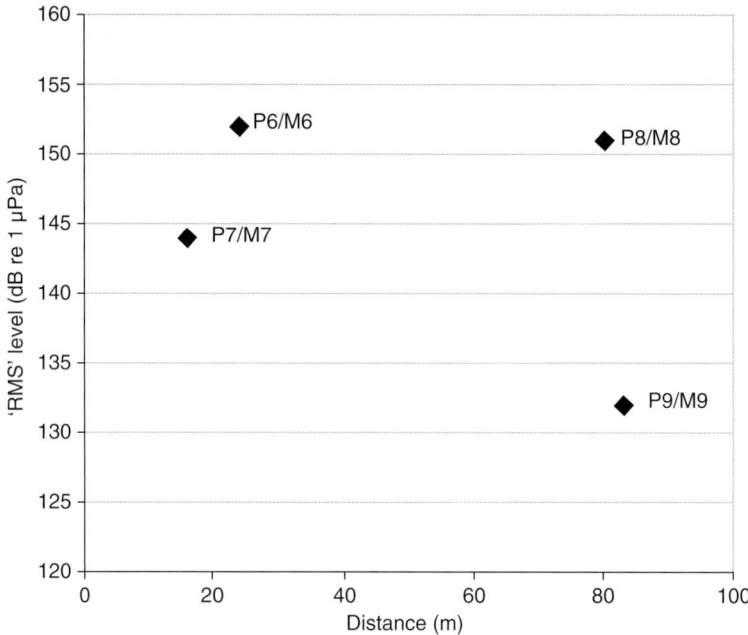

Fig. 9.12. The decline in sound pressure with distance from a vibratory pile driver. (From Nedwell and Edwards, 2002.[34])

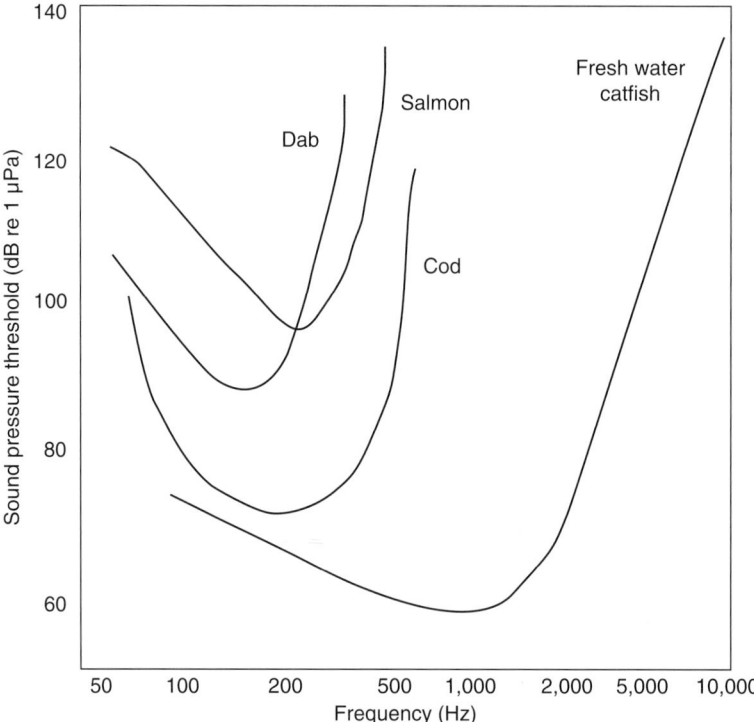

Fig. 9.13. Audiograms of teleost fish. (From Hawkins, 1986.[36])

The duration of the sound also has an effect on the response of fish, and short pulses of sound lasting only a few milliseconds may have a response threshold 30 or more decibels above that of a continuous sound (Hawkins, 1981).[38]

Salmon are only sensitive to low-frequency sound, measured in the low hundreds of Hz. This is at the lower end of sensitivity for birds and mammals and indicates that salmon are able to sense low-frequency vibrations but do not hear in the human sense. They detect particle motion rather than pressure change. The lowest response threshold and presumably the frequency of greatest sensitivity is between 100 and 160 Hz. Above 160 Hz sensitivity rapidly declines. A summary of the hearing sensitivity of salmonid and clupeid fish (shad are members of the herring family, which are clupeids) is given in Table 9.4.

Noise levels close to the pile driver (within about 5 m) can exceed the levels that will hurt or kill fish (peak quoted values are 218 dB). There are thus clear advantages in mitigating piling effects by scaring fish away from the pile before powerful pressure waves are generated. It is also important that there are periods of time over each day when no pile driving is undertaken to allow the fish to have unhindered passage.

Potential effects on fish of noise pollution caused by piling

Physical harm

Severe physical injuries resulting in the immediate death of fish occur when two critical criteria are met simultaneously (Larsen and Johnsen, 1992):[40]

Peak pressure >= 229 dB re 1 μPa and

Rise and decay time of =< 1 ms

Noise close to piling operations (within about 5 m) can hurt or kill fish (peak quoted values are around 218 dB), but it is unlikely that either impact or vibratory piling will cause instantaneous death because the pressure changes are insufficient even at close range.

Fish can suffer harm from which they may recover below 229 dB. Hastings (1990)[41] undertook studies for a military underwater acoustics project in a lake and observed that transient stunning occurred at 192–198 dB with frequencies of 150–400 Hz. Because stunned fish would usually be predated by other fish or birds, transient stunning should be considered as

Table 9.4. Optimum hearing frequencies, bandwidths and thresholds for herring and salmon. (Information supplied by Pisces Conservation Ltd.)

Species	Optimum frequency (Hz)	Bandwidth (Hz)	Auditory threshold (dB re 1 μP)	Reference
Herring (*Clupea harengus*)	300	20–800		Blaxter, 1981[38]
Salmon (*Salmo salar*)	160	100–220	99	Hawkins, 1986[35]

lethal. Hastings also found evidence of hearing damage in cod after exposure to 180 dB sound at frequencies of 50–400 Hz. He proposed that a safe sound level for lake fish was 150 dB.

The sound pressure at which harm occurs differs between species. Turnpenny et al. (1994)[42] noted that fish injury was related to the presence or absence, but not morphology, of their swimbladder. Fish can be divided into three groups: (i) those with no swimbladder; (ii) those with a swimbladder vented via a duct to the gut – physoclistous species; and (iii) those with an unvented swimbladder – physostomous species. In a series of experiments using frequencies between 95 and 1580 Hz, the effects of the presence of a swimbladder was clear. Sole, a flatfish species with no swim bladder, was uninjured at exposures of 177 dB. Bass and whiting, which are physoclistous species, showed no immediate signs of injury at 176 dB, although half the whiting died within 24 h. Trout, which are a physostomous species, were the most sensitive, with 57% mortality within 24 h among fish exposed to pressures of more than 170 dB at frequencies <500 Hz. Turnpenny et al. concluded that whether the swim bladder was vented conferred little advantage because the gas venting mechanism was unable to react quickly enough to the pressure changes.

Figure 9.14 summarizes present knowledge on the threshold sound levels at which fish experience physical injury.

Potential effects on diving mammals of noise pollution caused by piling

Pile driving has been observed to cause significant avoidance behaviour in marine mammals (e.g. Carstensen et al., 2006 and Dähne et al., 2013).[43, 44] Diving mammal densities can be reduced for a surprisingly large distance from a piling site. For example, Dähne et al. report on the effects

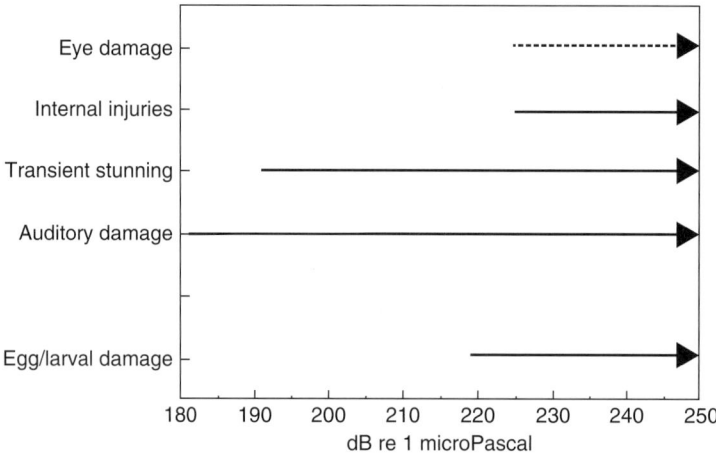

Fig. 9.14. The damage threshold for fish subjected to load sounds. (From Turnpenny et al., 1994.[42])

of percussive piling for 12 turbines in the Alpha Ventus wind farm in the German North Sea approximately 45 km north off the German coast. They observed a reduction in harbour porpoise observations at distances of up to 10.8 km from the site, with increased observations 25 to 50 km distant. These results were interpreted as a displacement caused by piling.

Notes

[1] IPCC (2011) *IPCC Special Report on Renewable Energy Sources and Climate Change Mitigation*. Prepared by Working Group III of the Intergovernmental Panel on Climate Change [O. Edenhofer, R. Pichs-Madruga, Y. Sokona, K. Seyboth, P. Matschoss, S. Kadner, T. Zwickel, P. Eickemeier, G. Hansen, S. Schlömer, C. von Stechow (eds)]. Cambridge University Press, Cambridge.

[2] Bergström, L., Kautsky, L., Malm, T., Rosenberg, R., Wahlberg, M., Capetillo, N.Å. and Wilhelmsson, D. (2014) Effects of offshore wind farms on marine wildlife – a generalized impact assessment. *Environmental Research Letters* 9(3), 034012.

[3] Nedwell, J. and Howell, D. (2004) *A review of offshore windfarm related underwater noise sources*. Report No. 544 R 0308, p. 63. Available at: www.subacoustech.com/wp-content/uploads/544R0308.pdf (accessed 3 April 2018).

[4] Fawley Aquatic Research Laboratories (FARL) (1995) *Possible impacts of dredging on salmonids*. Research Note for ABP Research. Fawley Aquatic Research Laboratories Ltd, Southampton, UK.

[5] Clarke, D., Engler, R.M. and Wilber, D.H. (2000) *Assessment of Potential Impacts of Dredging Operations Due to Sediment Resuspension* (No. ERDC-TN-DOER-E9). Army Engineer Waterways Experiment Station, Vicksburg, USA. Available at: https://dots.el.erdc.dren.mil/doer/pdf/doere9.pdf (accessed 7 December 2017).

[6] Kiørboe, T., Frantsen, E., Jensen, C. and Sørensen, G. (1981) Effects of suspended sediment on development and hatching of herring (*Clupea harengus*) eggs. *Estuarine, Coastal, and Shelf Science* 13(1), 107–111.

[7] Messieh, S.N., Wildish, D.J. *et al.* (1981) *Possible impact from dredging and spoil disposal on the Miramichi Bay herring fishery*. Canadian Technical Report on Fisheries and Aquatic Science.

[8] McNair, E.C.J. and Banks, G.E. (1986) Prediction of flow fields near the suction of a cutterhead dredge. *American Malacological Bulletin* Special Edition No. 3, 37–40.

[9] Armstrong, D., Stevens, B. *et al.* (1982) *Distribution and abundance of Dungeness crab and Crangon shrimp, and dredged-related mortality of invertebrates and fish in Grays Harbor, Washington*. School of Fisheries, University of Washington, Washington Department of Fisheries, and U.S. Army Engineer District, Seattle, Washington.

[10] Reine, K. and Clarke, D. (1998) Entrainment by hydraulic dredges – a review of potential impacts. U.S. Army Engineer Research and Development Center, Vicksburg, Mississippi.

[11] Wilhelmsson, D., Malm, T. and Öhman, M.C. (2006) The influence of offshore windpower on demersal fish. *ICES Journal of Marine Science* 63(5), 775–784.

[12] Bulleri, F. and Airoldi, L. (2005) Artificial marine structures facilitate the spread of a nonindigenous green alga, *Codium fragile* ssp. *tomentosoides*, in the north Adriatic Sea. *Journal of Applied Ecology* 42(6), 1063–1072.

[13] Reubens, J.T., Degraer, S. and Vincx, M. (2011) Aggregation and feeding behaviour of pouting (*Trisopterus luscus*) at wind turbines in the Belgian part of the North Sea. *Fisheries Research* 108(1), 223–227.

[14] Lindeboom, H.J., Kouwenhoven, H.J., Bergman, M.J.N., Bouma, S., Brasseur, S.M.J.M. *et al.* (2011) Short-term ecological effects of an offshore wind farm in the Dutch coastal zone; a compilation. *Environmental Research Letters* 6(3), 035101.

[15] Gill, A.B., Bartlett, M. and Thomsen, F. (2012) Potential interactions between diadromous fishes of UK conservation importance and the electromagnetic fields and subsea noise from marine renewable energy developments. *Journal of Fish Biology* 81(2), 664–695.

[16] Bergström, L., Kautsky, L., Malm, T., Rosenberg, R., Wahlberg, M., Capetillo, N.Å. and Wilhelmsson, D. (2014) Effects of offshore wind farms on marine wildlife – a generalized impact assessment. *Environmental Research Letters* 9(3), 034012.

[17] Hüppop, O., Dierschke, J., Exo, K.M., Fredrich, E. and Hill, R. (2006) Bird migration studies and potential collision risk with offshore wind turbines. *Ibis* 148(s1), 90–109.

[18] Rydell, J., Bach, L., Bach, P., Diaz, L.G., Furmankiewicz, J. *et al.* (2014) Phenology of migratory bat activity across the Baltic Sea and the south-eastern North Sea. *Acta Chiropterologica* 16(1), 139–147.

[19] Johnson, J.B., Gates, J.E. and Zegre, N.P. (2011) Monitoring seasonal bat activity on a coastal barrier island in Maryland, USA. *Environmental Monitoring and Assessment* 173(1–4), 685–699.

[20] Denholm, P., Hand, M., Jackson, M. and Ong, S. (2009) *Land-use Requirements of Modern Wind Power Plants in the United States.* National Renewable Energy Laboratory, Golden, Colorado, USA.

[21] RSPB (n.d.) *Wind Farms.* Available at: https://www.rspb.org.uk/our-work/our-positions-and-casework/our-positions/climate-change/action-to-tackle-climate-change/uk-energy-policy/wind-farms (accessed 7 December 2017).

[22] Drewitt, A.L. and Langston, R.H. (2006) Assessing the impacts of wind farms on birds. *Ibis* 148(s1), 29–42.

[23] Pearce-Higgins, J.W., Stephen, L., Langston, R.H., Bainbridge, I.P. and Bullman, R. (2009) The distribution of breeding birds around upland wind farms. *Journal of Applied Ecology* 46(6), 1323–1331.

[24] Hötker, H., Thomsen, K.M. and Köster, H. (2006) *Impacts on biodiversity of exploitation of renewable energy sources: the example of birds and bats. Facts, gaps in knowledge, demands for further research, and ornithological guidelines for the development of renewable energy exploitation.* Michael-Otto-Institut im NABU, Bergenhusen.

[25] Smallwood, K.S. (2013) Comparing bird and bat fatality-rate estimates among North American wind-energy projects. *Wildlife Society Bulletin* 37(1), 19–33.

[26] Farfán, M.A., Duarte, J., Real, R., Muñoz, A.R., Fa, J.E. and Vargas, J.M. (2017) Differential recovery of habitat use by birds after wind farm installation: a multi-year comparison. *Environmental Impact Assessment Review* 64, 8–15.

[27] Arnett, E.B., Baerwald, E.F., Mathews, F., Rodrigues, L., Rodríguez-Durán, A. *et al.* (2016) Impacts of wind energy development on bats: a global perspective. In: Voigt, C.C. and Kingston, T. (eds) *Bats in the Anthropocene: Conservation of Bats in a Changing World.* Springer International, Cham, Switzerland, pp. 295–323.

[28] Kunz, T.H., Arnett, E.B., Erickson, W.P., Hoar, A.R., Johnson, G.D., Larkin, R.P., Strickland, M.D., Thresher, R.W. and Tuttle, M.D. (2007) Ecological impacts of wind energy development on bats: questions, research needs, and hypotheses. *Frontiers in Ecology and the Environment* 5(6), 315–324.

[29] Cryan, P.M. and Barclay, R.M. (2009) Causes of bat fatalities at wind turbines: hypotheses and predictions. *Journal of Mammalogy* 90(6), 1330–1340.

[30] Corten, G.P. and Veldkamp, H.F. (2001) Aerodynamics: insects can halve wind-turbine power. *Nature* 412(6842), 41.

[31] Long, C.V., Flint, J.A. and Lepper, P.A. (2011) Insect attraction to wind turbines: does colour play a role? *European Journal of Wildlife Research* 57(2), 323–331.

[32] Liu, P. and Barlow, C.Y. (2017) Wind turbine blade waste in 2050. *Waste Management* 62, 229–240.

[33] Nedwell, J., Turnpenny, A., Langworthy, J. and Edwards, B. (2003) *Measurements of underwater noise during piling at the Red Funnel Terminal, Southampton, and observations of its effect on caged fish.* Report for Red Funnel. Report Reference: 558 R 0207.

[34] Nedwell, J. and Edwards, B. (2002) *Measurements of underwater noise in the Arun River during piling at County Wharf, Littlehampton.* Report for David Wilson Homes Ltd., Report Reference: 513 R 0108.

[35] Nedwell, J., Turnpenny, A., Langworthy, J. and Edwards, B. (2003) *Measurements of underwater noise during piling at the Red Funnel Terminal, Southampton, and observations of its effect on caged fish.* Report for Red Funnel. Report Reference: 558 R 0207

[36] Hawkins, A.D. (1986) Underwater sound and fish behaviour. In: Pitcher, T.J. (ed.) *The Behaviour of Teleost Fishes.* Chapman & Hall, London, pp. 129–170.

[37] Buerkle, K. (1968) Relation of pure tone thresholds to background noise level in the Atlantic cod (*Gadus morhua*). *Journal of the Fisheries Research Board of Canada* 25(6), 1155–1160.

[38] Hawkins, A.D. (1981) The hearing abilities of fish. In: Tavolga, W.N., Popper, A.N. and Fay, R.R. (eds) *Hearing and Sound Communication in Fishes.* Springer, New York, pp. 109–138.

[39] Blaxter, J.H.S. (1981) The swimbladder and hearing. In: Tavolga, W.N., Popper, A.N. and Fay, R.R. (eds) *Hearing and Sound Communication in Fishes.* Springer, New York, pp. 61–71.

[40] Larsen, T. and Johnsen, H.K. (1992) Recent experience on the impact of chemical explosives on penned Atlantic salmon and cod. *Fisheries and Offshore Petroleum Exploitation 2nd International Conference*, Bergen, Norway.

[41] Hastings, M.C. (1990) *Effects of Underwater Sound on Fish.* Document No. 46254-900206-01IM, Project No. 401775-1600, AT&T Bell Laboratories.

[42] Turnpenny, A.W.H., Thatcher, K.P. and Nedwell, J.R. (1994) *The effects on fish and other marine animals of high-level underwater sound.* Report FRR 127/94, Fawley Aquatic Research Laboratories, Ltd., Southampton, UK.

[43] Carstensen, J., Henriksen, O.D. and Teilmann, J. (2006) Impacts of offshore wind farm construction on harbour porpoises: acoustic monitoring of echolocation activity using porpoise detectors (T-PODs). *Marine Ecology Progress Series* 321, 295–308.

[44] Dähne, M., Gilles, A., Lucke, K., Peschko, V., Adler, S. *et al.* (2013) Effects of pile-driving on harbour porpoises (*Phocoena phocoena*) at the first offshore wind farm in Germany. *Environmental Research Letters* 8(2), 025002.

10 Solar Power

Solar generators can be divided into two broad types, photovoltaic (PV) solar cells and concentrating solar power (CSP) plants. The scale of PV generation varies greatly from single panel systems to recharge a battery via individual house rooftop PV arrays (Fig. 10.1) to large utility-scale PV and CSP projects. The ecological impacts of solar generation have been reviewed by Tsoutsos *et al.* (2005).[1] As the installed capacity increases and operational experience expands, it is likely that ecological impacts and concerns will also increase.

Photovoltaic Generation

We are all becoming familiar with solar panels fitted on our houses, farm outbuildings, commercial buildings and boats. There is also a steadily growing number of larger-scale facilities. Photovoltaic cells use light to generate electricity. A variety of panel designs are available, including polycrystalline, monocrystalline and thin-film. The technology is rapidly developing and new designs are continually being tested. It is now possible to install PV roof tiles.

Growth in installed PV capacity

Worldwide installed PV capacity has grown at an explosive rate since 2005 (Fig. 10.2). Over the year 2016 to 2017, capacity was increasing at a rate that would generate a doubling in capacity about every 2 years. As shown in Table 10.1, China now has the greatest installed capacity and is installing new capacity at the fastest rate. Although slow to commence large-scale

Fig. 10.1. A typical roof-mounted domestic solar panel array. (Photo courtesy of Pujanak, public domain via Wikimedia Commons.)

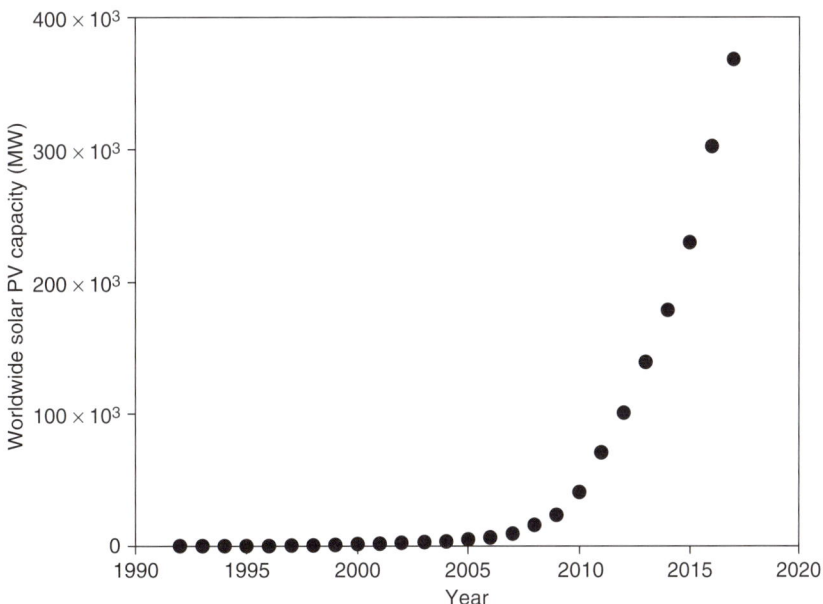

Fig. 10.2. The growth in worldwide solar capacity. (Data for 1996–1999: BP-Statistical Review of world energy (Historical Data Workbook); for 2000–2013: EPIA Global Outlook on Photovoltaics Report. Recent values are estimates.)

installation, India has now constructed one of the world's largest solar farms, The Kurnool Solar Park, located over a 24 km² area in the Gani and Sakunala villages within the arid Kurnool district, Andhra Pradesh. The park utilizes over 4 million solar panels with capacities of 315 and 320 W. Installed capacity in the USA is also now rapidly increasing and

Table 10.1. The installed PV capacity in 2016 for the 30 countries with the greatest installed capacity. (Data from http://www.iea-pvps.org/fileadmin/dam/public/report/PICS/IEA-PVPS_-__A_Snapshot_of_Global_PV_-_1992-2015_-_Final_2_02.pdf.)

	Country	Total installed capacity (MW)	Capacity installed 2015–2016 (MW)
1	China	78,070	34,540
2	Japan	42,750	8,600
3	Germany	41,220	1,520
4	USA	40,300	14,730
5	Italy	19,279	373
6	United Kingdom	11,630	1,970
7	India	9,010	3,970
8	France	7,130	559
9	Australia	5,900	839
10	Spain	5,490	55
11	South Korea	4,350	850
12	Belgium	3,422	170
13	Canada	2,715	200
14	Thailand	2,150	726
15	Netherlands	2,100	525
16	Switzerland	1,640	250
17	Chile	1,610	746
18	South Africa	1,450	536
19	Austria	1,077	154
20	Israel	910	130
21	Philippines	900	756
22	Denmark	900	70
23	Turkey	832	584
24	Portugal	513	58
25	Mexico	320	150
27	Malaysia	286	54
28	Sweden	175	60
29	Norway	26.7	11
30	Finland	15	10

it is likely to radically disrupt the economics of utilities running conventional steam plants. Progress is now so rapid that it is inevitable that any figures presented here will be out of date by the time of publication.

Ecological impacts of large-scale solar

Ecological impacts relating to large-scale solar power generation are discussed by Turney and Fthenakis (2011).[2] The areas of concern are discussed below.

Land use

Large-scale PV sites occupy considerable areas of land. For example, the Topaz Solar Farm is a 550-MW photovoltaic power station in San Luis Obispo

County, California. It occupies a 9.5 mile^2 (25 km^2) site (see Figs 10.3 and 10.4). Topaz comprises 9 million CdTe photovoltaic modules and is visible from space. Depending on their location, larger utility-scale solar PV facilities can raise concerns about land degradation and habitat loss. Total land area requirements vary with technology, the topography of the site and the intensity of the solar resource. Estimates of land required for utility-scale PV systems range from 3.5 to 10 acres/MW (14,164 to 40,468 m^2/MW). As an example of the land requirement for a wet northern temperate country, within Britain a 1-MW installation requires 5–6 acres and would possibly generate 800,000–900,000 kWh/year/MW of installed capacity. This will vary with location, however. In Britain, large-scale PV developments favour brown-field or low-grade agricultural land, although they do occur on agricultural land of better quality and typically preserve the original field layout and hedges. It should be possible to remove PV panels and quickly convert land to other uses, including agriculture. However, the effects of long-term shading on soil characteristics are unclear.

While countries with large deserts have the opportunity to use low biomass habitat with high-intensity solar inputs (e.g. Longyangxia Dam Solar Park, Fig. 10.5), such farms will inevitably be in regions with low population densities. Unlike wind turbines, there is less opportunity for solar projects to share land with agricultural uses. However, land impacts from utility-scale solar systems can be minimized by siting them at lower-quality locations such as brown-field sites, abandoned mining land or existing transportation and transmission corridors.

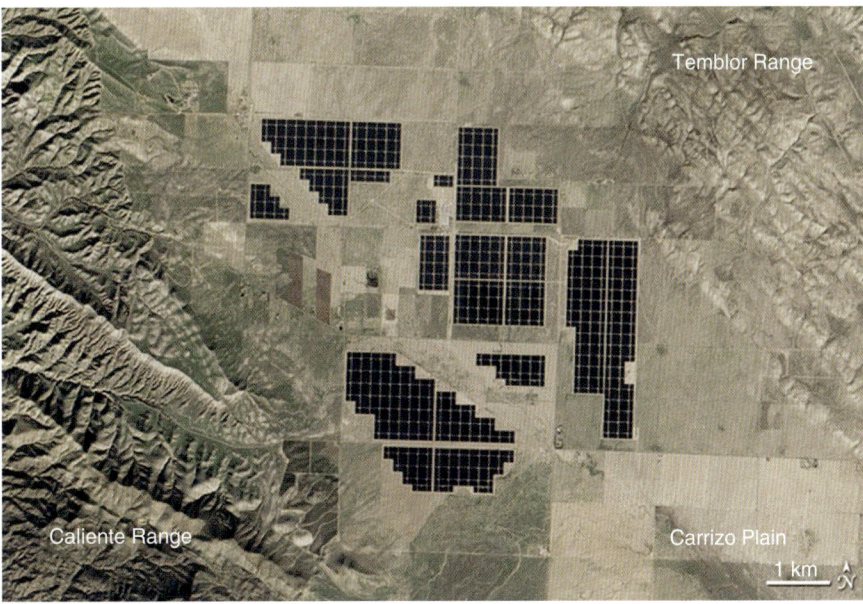

Fig. 10.3. Topaz Solar Farm from space. (2015 Earth Observatory image by Jesse Allen, using EO-1 ALI data provided courtesy of the NASA EO-1 team.)

Fig. 10.4. Topaz Solar Farm panels. (Photo courtesy of Sarah Swenty/USFWS.)

Large-scale PV facilities have a visual impact, the importance of which will vary with the design, topography of the site and the views and tastes of the local residents.

Water use
Solar PV cells do not use water for generating electricity. However, as in all manufacturing processes, some water is used to manufacture solar PV components. In some regions dust and other debris must be regularly washed off panels. The water required for this is unknown, but unlikely to be great.

Release of pollutants during operation
During normal operation, PV systems emit no gaseous, liquid or radio-active pollutants. CIS and CdTe modules include small quantities of toxic substances, so there is a potential risk that a fire in an array might release these chemicals to the environment. At large-scale PV facilities the acci-dental release of hazardous materials with subsequent risk to the public and workers needs to be anticipated and an emergency response devel-oped to limit exposure and emissions to soil and groundwater.

Impacts on animals and plants
The environmental impact report[3] prepared for the 550 MWp Topaz photo-voltaic project (see Figs 10.3 and 10.4) in grasslands and abandoned farm-lands of central California identified the potential for significant impacts on many protected animal and plant species. Using mitigation measures funded by the project, it was claimed these impacts were made insignificant.

Fig. 10.5. Longyangxia Dam Solar Park, which at 850-MW installed capacity was the world's largest solar farm in 2017. (Photo courtesy of USGS/NASA Landsat.)

We will need more operational experience before a full assessment of the impacts can be reached, because as Turney and Fthenakis (2011)[4] state, 'it should be kept in mind that monitoring of impacts is just beginning.'

Possible mitigation approaches that can be offered include:

- Elimination of invasive species.
- Control of damaging species, such as deer or rabbits.
- Construction of suitable habitat for endemic species.
- The exclusion of man and his pets from sensitive habitats, e.g. reduced access to recreational off-road vehicles and dogs.
- Increased monitoring of the ecosystem.

There is an appreciable risk to flying animals from accidental impacts. Kagan *et al.* (2014)[5] compared bird deaths at three solar energy facilities in southern California: Desert Sunlight (photovoltaic), Genesis (trough system concentrated solar) and Ivanpah (power tower concentrated solar). While there has been considerable coverage of the fact that concentrated solar facilities can burn birds, it is notable that they observed appreciable numbers of deaths at the Desert Sunlight photovoltaic facility. These were caused by impact trauma and predation. That glass panels can present a hazard to birds is well established. Sheet glass in buildings is estimated to kill between 365 and 988 million birds annually in the USA alone (Loss *et al.* 2014).[6]

Kagan *et al.* (2014) note that in a desert environment a large expanse of reflective, blue panels may appear like a large body of water, causing birds to attempt a landing. They note that birds for which the primary habitat is water, including coots, grebes and cormorants, were over-represented in mortalities at the Desert Sunlight facility (44% of the total) compared with the CSP plants of Genesis (19%) and Ivanpah (10%). A problem for some aquatic birds like grebes is that once landed on or near a panel, even if they avoid impact trauma, they experience difficulty in taking off and so are vulnerable to predators. They suggest that panels be modified to have visual cues to make them different from water, each panel should have UV-reflective or solid, contrasting bands spaced no further than 28 cm apart.

The attraction of aquatic insects is discussed below under small-scale PV installations.

Hazardous materials

The PV cell manufacturing process uses a number of hazardous materials. These include hydrochloric acid, sulphuric acid, nitric acid, hydrogen fluoride, 1,1,1-trichloroethane and acetone. The amount and type of chemicals used depends on the type of cell, the amount of cleaning that is needed and the size of silicon wafer. Workers also face risks associated with inhaling silicon dust.

Thin-film PV cells contain a higher number of more toxic materials than those used in traditional silicon photovoltaic cells, including gallium arsenide, copper-indium-gallium-diselenide and cadmium-telluride. If not handled and disposed of properly, these materials could pose serious environmental or public health threats. However, manufacturers have a strong financial incentive to ensure that these expensive and scarce materials are recycled rather than thrown away and good recycling schemes are in place. For example, First Solar has a process for recycling CdTe thin-film modules. This can be summarized as follows (see http://www.renewableenergyfocus. com/view/3005/end-of-life-pv-then-what-recycling-solar-pv-panels/):

- The costs of collection and transportation to the recycling centre are built into the module sale price.
- Modules are shredded into large pieces and crushed by a hammer mill to pieces typically smaller than 5 mm in order to break lamination bonds.

- Semiconductor films are removed in a slow rating leach drum in a process taking 4–6 h. Weak sulphuric acid and hydrogen peroxide is added to the glass to achieve an optimal solid–liquid ratio. The films are etched from the glass during the leach cycle.
- Glass is separated from the liquids.
- A vibrating screen separates glass from the larger ethylene vinyl acetate pieces.
- The glass is sent for recycling and the rinse waters are pumped to a precipitation system for metal recovery.
- The metal compounds are precipitated and formed into a metal-rich filter cake, which is sent for processing to semiconductor grade raw materials for use in new solar modules.

It is claimed the process can recover 90% of the glass and 95% of the semiconductor materials. According to BNL, the recovery of tellurium is 80% or better and it can be sold commercially.

Life-cycle carbon emissions
While there are no global warming emissions associated with generating electricity from solar energy, there are emissions associated with other stages of the solar panel construction and disposal cycle, including manufacturing, materials transportation, installation, maintenance, and decommissioning and dismantling. Most estimates of life-cycle emissions for photovoltaic systems are between 0.08 and 0.2 pounds of carbon dioxide equivalent per kilowatt-hour. As a comparison, lifecycle emission rates for natural gas are estimated as 0.6–2 lb of CO_2e/kWh and coal 1.4–3.6 lb of CO_2e/kWh.[7]

Ecological impacts of small-scale PV arrays

Small-scale solar PV arrays are now commonly placed on homes and commercial buildings where they have minimal land-use impact. Small, 1–4 kW capacity solar panel arrays for domestic households typically export excess power to the local grid.

Construction and recycling issues are as for large-scale PV facilities, although with the added problems linked to dispersal, so it is difficult to ensure good practice. It is inevitable that, with time, broken and abandoned panels will accumulate on and within badly maintained buildings and gardens.

Harm to flying insects
It is notable that some flying insects collide with and are injured by PV panels. The extent of this issue is unknown. It is possible that aquatic insects, such as water beetles, may mistake the reflective surface for water and attempt to dive. For example, Kriska *et al.* (2008)[8] found that vertically oriented black glass surfaces (similar to solar panels) produced highly

polarized reflected light that makes them highly attractive to aquatic insects. It would be interesting to see the results of studies comparing the insect fauna living close to a PV array with those of an unaffected control area. Given the millions of domestic PV arrays now installed in Europe and North America, aquatic insects are probably more likely to encounter a PV panel than a small pond.

Concentrating Solar Thermal Power Plants

Concentrated solar thermal systems generate power using mirrors or lenses to concentrate sunlight onto a small area to generate heat to drive a steam turbine or other heat engine, which is used to generate electricity (see Figs 10.6 and 10.7). In some designs heat is stored in molten salts so that electricity can be generated when sunlight is unavailable.

There are a number of different designs, including:

- Parabolic trough.
- Enclosed trough.
- Solar power tower.
- Fresnel reflectors.
- Dish Stirling.

Fig. 10.6. Crescent Dunes Solar Energy Project, as seen from an airliner. It is a 110-MW net solar thermal power project with 1.1 gigawatt-hours of energy storage, located near Tonopah, about 190 miles (310 km) northwest of Las Vegas. (Photo courtesy of Amble, published under a CC BY-SA 4.0 license via Wikimedia Commons.)

Fig. 10.7. Solucar PS10. The first tower-based commercial solar thermal power plant, Andalucía, Spain. (Photo courtesy of afloresm, published under a CC BY 2.0 license via Wikimedia Commons.)

Concentrating solar power (CSP) has a comparatively long history for an alternative technology. The first CSP plant was completed near Genoa, Italy in 1968. The 10-MW Solar One power tower was built in southern California in 1981, followed by a parabolic-trough design at the nearby Solar Energy Generating Systems in 1984. At 354 MW, this plant remains the largest solar power plant presently in operation.

Until recently, CSP facilities were concentrated in Spain and the USA. Spain is the world leader in CSP with an installed capacity in January 2016 of 2300 MW, followed by the USA with 1740 MW. Global capacity is dominated by parabolic-trough plants, which account for 90% of CSP plants. By late 2016, facilities were under construction in Australia, Chile, China, India, Israel, Mexico, Saudi Arabia and South Africa. Morocco has become a major constructor of CSP plants: the 500-MW Noor-Ouarzazate CSP complex is expected to be fully operational by 2018. As shown in Fig. 10.8, installed capacity is now beginning to increase rapidly, although this may not be sustained given the present cost advantages of photovoltaic arrays.

In most cases, CSP technologies cannot presently compete on price with photovoltaics (solar panels), which have experienced huge growth in recent years due to falling prices of the panels and much smaller operating costs. Recently built facilities incorporate thermal energy storage capacity, which is seen as central to maintaining the competitiveness of CSP because it allows electricity generation when the sun is not shining.

Many of the regions of our planet that have the highest potential for solar energy also tend to be those with the driest climates, so careful consideration of these water trade-offs is essential.

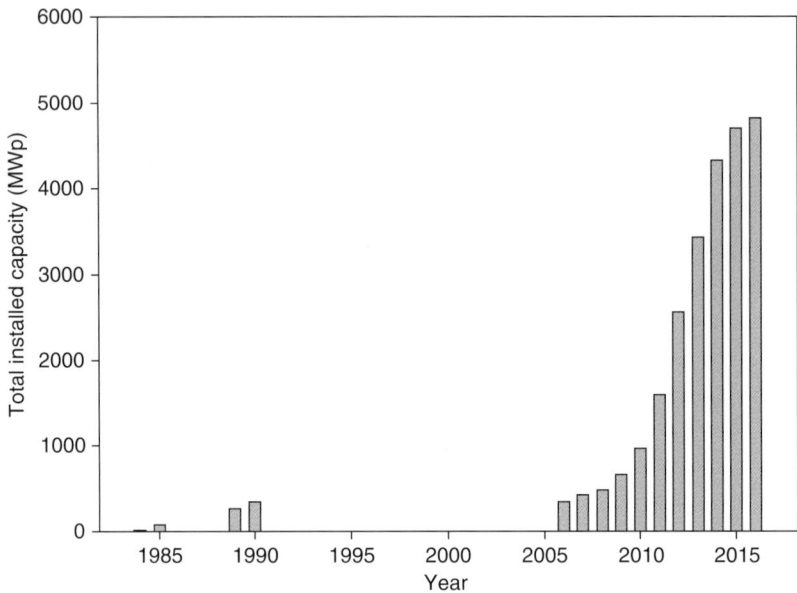

Fig. 10.8. The growth in worldwide concentrating solar power plant capacity. (From International Renewable Energy Agency[9] and Renewable Energy Policy Network for the 21st Century[10] – http://www.ren21.net/.)

Ecological impacts of concentrating solar power plants

Given the small number of plants and the limited operational period, we do not have sufficient data on which to make a full assessment of ecological impacts. During operation they produce no emissions. However, there will be appreciable emissions linked to construction and manufacture of components.

Water use and thermal pollution

Concentrating solar power (CSP) plants, like all thermal electric plants, require water for cooling. Water use depends on the plant design, plant location and the type of cooling system. CSP plants that use wet-recirculating technology with cooling towers withdraw between 600 and 650 gallons of water per megawatt-hour of electricity produced. CSP plants with once-through cooling technology have higher levels of water withdrawal, but lower total water consumption (because water is not lost as steam). Dry-cooling technology can reduce water use at CSP plants by approximately 90%. However, the trade-offs to these water savings are higher costs and lower efficiencies. In addition, dry-cooling technology is significantly less effective at temperatures above 100°F. The effects of thermal pollution and water use in cooling towers are discussed in Chapter 5 (this volume).

Impacts on flying organisms

Concentrating solar power systems pose a danger to birds, bats and insects. The literature is confusing as to the extent of the problem. Tsoutsos *et al.*

(2005)[11] indicate that operational experience shows that birds avoid any danger areas and that the losses to insect populations are insignificant. However, there have been a series of reports in the US press of scorched birds[12] that were hunting insects attracted to the bright light. This impact has passed further up the food chain because it has also affected the raptors who hunt the birds. The Associated Press reported that birds were being killed by concentrated sunlight at a rate of one bird every 2 minutes (28,000 per year) at the Ivanpah plant, California, which created considerable concern. Within the USA in particular, there has been widespread speculation and wild estimation of the actual number of birds and other organisms killed. Mr Holland of NRG Energy, the part owner together with BrightSource Energy and Google, states that 'Ivanpah reported 321 total avian fatalities between January and June 2014, of which 133 were related to sunlight being reflected onto the boilers, thus falling far short of the "estimates" in the Associated Press article'. It has been estimated from the sampling data that this plant produces about 1000 bird deaths per year.[13] The industry believes that there are means of reducing avian mortality by redirecting mirrors when not needed.

Ho (2016)[14] presents an analysis of bird deaths at CSP plants using data collected at Solar One in California, the Solar Energy Development Center in Israel, Ivanpah Solar Electric Generating System (ISEGS) in California, Crescent Dunes in Nevada and Gemasolar in Spain. He notes that safe irradiance levels for birds have been reported to range between 4 and 50 kW/m^2. Above these levels, singeing and irreversible damage to the feathers occur (see Fig. 10.9). The most distressing observation is large numbers of 'streamers' in high solar flux areas, which are presumably burning, possibly completely vaporizing, animals. Ho concludes that

Fig. 10.9. Warbler burned mid-air by solar thermal power plant. (Photo courtesy of US Fish and Wildlife Service, public domain via Wikimedia Commons.)

these streamers are unlikely to represent completely vaporized birds and in most cases are insects. However, this conclusion has been questioned by Kagan *et al.* (2014).[16] They observed many streamer events, for some of which the amount of smoke produced by the ignition could only be explained by a larger flammable biomass such as a bird. Staff actually observed birds entering the solar flux and igniting, consequently becoming a streamer, suggesting that while complete vaporization may be rare, crashing to the ground in flames may be unacceptably common. A video of combusting animals is available at https://www.youtube.com/watch?v=ICLXQN_lURk&feature=youtu.be.

Ho (2016) presents an estimated standardized bird mortality rate during the first year of operation at Ivanpah as 0.7–3.5 fatalities per GWh. It should be noted, however, that this is just the mortality linked to concentrated solar flux and there are also collision mortalities (see also the discussion of impacts in the PV section above). Kagan *et al.* (2014) considered the actual cause of bird death and noted five potential pathways:

- trauma from collision after losing the ability to fly as a result of direct heat damage to feathers;
- starvation and/or thermoregulatory dysfunction after direct heat damage to feathers;
- shock;
- soft tissue damage after whole-body exposure to high heat; and
- ocular damage after exposure to bright light.

They concluded from the study of corpses that the evidence favoured the first three mechanisms.

McCrary *et al.* (1986)[15], in a study of bird deaths at the Solar One 10 MWe direct-steam pilot demonstration project, estimated a mortality rate of 1.9–2.2 birds per week. Of the 70 documented bird deaths, 57 (81%) died from collisions with Solar One structures (mainly heliostats) and 13 (19%) died from burns sustained while flying through high-flux space. This suggests that burning is only part of the problem.

Kagan *et al.* (2014)[16] argue that solar flux facilities may act as 'mega-traps', which they define as artificial features that attract and kill species at multiple trophic levels. The strong light emitted by these facilities attracts insects, which in turn attract insect-eating birds, which are incapacitated by solar flux injury, thus attracting predators and creating an entire food chain vulnerable to injury and death. They observed at Ivanpah 'hundreds upon hundreds of butterflies (including monarchs, *Danaus plexippus*) and dragonfly carcasses. Some showed singeing, and many appeared to have just fallen from the sky'. Observations showed the insects to be flying in the bright area around the boiler at the top of the tower and the insects were attracted to the light. Birds were also observed feeding on the insects. Insect aggregations may also attract bats, which were observed roosting at the base of the Ivanpah tower.

Kagan *et al.* (2014) suggest the following steps could be taken at the Ivanpah plant to reduce bird and bat deaths:

- Clear vegetation from the ground around the plant to reduce the attractiveness of the habitat to birds.
- Fit UV reflective strips to the solar panels to give birds a visual cue.
- Suspend power generation seasonally when vulnerable species are migrating.
- Avoid rotating the mirrors into a vertical position.
- Net over ponds.
- Place perch deterrent devices on structures near to high-energy flux areas.
- Stop bats roosting near the condensers.

It is clear that considerably more information is needed on the impacts of CSP plants on flying animals of all kinds. As with wind turbines, video cameras and possibly specialized radar should be used to fully understand the extent of the problem. Much more effort needs to be focused on insect losses at all solar generating facilities.

Health and safety (occupational hazards)

The release of heat transfer fluids (water and oil) from parabolic trough and central receiver systems could be hazardous. Central tower systems, which use liquid sodium or molten salts as a heat-transfer medium, offer particular hazards and a fatal accident has occurred in a system using liquid sodium.[17] Central tower systems have the potential to concentrate light to intensities that could damage eyesight. Under normal operating conditions this will not pose a danger to operators, but failure of the tracking systems could result in straying beams that might present a hazard.

Notes

[1] Tsoutsos, T., Frantzeskaki, N. and Gekas, V. (2005) Environmental impacts from the solar energy technologies. *Energy Policy* 33(3), 289–296.

[2] Turney, D. and Fthenakis, V. (2011) Environmental impacts from the installation and operation of large-scale solar power plants. *Renewable and Sustainable Energy Reviews* 15(6), 3261–3270.

[3] San Luis Obispo Department Planning and Building (2010) *Draft Environmental Impact Report for the Topaz Solar Project.* County of San Luis Obispo Planning Department, San Luis Obispo County, California.

[4] Turney, D. and Fthenakis, V. (2011) Environmental impacts from the installation and operation of large-scale solar power plants. *Renewable and Sustainable Energy Reviews* 15(6), 3261–3270.

[5] Kagan, R.A., Viner, T.C., Trail, P.W. and Espinoza, E.O. (2014) *Avian Mortality at Solar Energy Facilities in Southern California: A Preliminary Analysis.* National Fish and Wildlife Forensics Laboratory, Ashland, Oregon, USA.

[6] Loss, S.R., Will, T., Loss, S.S. and Marra, P.P. (2014) Bird–building collisions in the United States: estimates of annual mortality and species vulnerability. *The Condor* 116(1), 8–23.

[7] IPCC (2011) *IPCC Special Report on Renewable Energy Sources and Climate Change Mitigation.* Prepared by Working Group III of the Intergovernmental Panel on Climate Change [Edenhofer, O., Pichs-Madruga, R., Sokona, Y., Seyboth, K., Matschoss, P., Kadner, S., Zwickel, T., Eickemeier, P., Hansen, G., Schlömer, S., von Stechow, C. (eds)]. Cambridge University Press, Cambridge, UK, Chapters 7 and 9.

[8] Kriska, G., Malik, P., Szivák, I. and Horváth, G. (2008) Glass buildings on river banks as 'polarized light traps' for mass-swarming polarotactic caddis flies. *Naturwissenschaften* 95(5), 461–467.

[9] International Renewable Energy Agency (2012) *Concentrating Solar Power.* International Renewable Energy Agency, Abu Dhabi, UAE.

[10] REN21 (2016) *Renewables 2016: Global Status Report.* Available at: http://www.ren21. net/wp-content/uploads/2016/06/GSR_2016_Full_Report.pdf (accessed 8 December 2017).

[11] Tsoutsos, T., Frantzeskaki, N. and Gekas, V. (2005) Environmental impacts from the solar energy technologies. *Energy Policy* 33(3), 289–296.

[12] NBC News (2014) *Burned Birds Become New Environmental Victims of the Energy Quest.* Available at: http://www.nbcnews.com/science/environment/burned-birds-become-new-environmental-victims-energy-quest-n184426 (accessed 8 December 2017).

[13] H.T. Harvey & Associates (2014) *Ivanpah Solar Electric Generating System Avian & Bat Monitoring Plan.* Available at: http://docketpublic.energy.ca.gov/PublicDocuments/07-AFC-05C/TN202617_20140626T131025_Ivanpah_Solar_Electric_System_Avian__Bat_Monitoring_Plan;_20132.pdf (accessed 8 December 2017).

[14] Ho, C.K. (2016) Review of avian mortality studies at concentrating solar power plants. *AIP Conference Proceedings* 1734(1), 070017.

[15] McCrary, M.D., McKernan, R.L., Schreiber, R.W., Wagner, W.D. and Sciarrotta, T.C. (1986) Avian mortality at a solar energy power plant. *Journal of Field Ornithology* 57(2), 135–141.

[16] Kagan, R.A., Viner, T.C., Trail, P.W. and Espinoza, E.O. (2014) *Avian Mortality at Solar Energy Facilities in Southern California: A Preliminary Analysis.* National Fish and Wildlife Forensics Laboratory, Ashland, Oregon.

[17] Tsoutsos, T., Frantzeskaki, N. and Gekas, V. (2005) Environmental impacts from the solar energy technologies. *Energy Policy* 33(3), 289–296.

11 Fuel Cells and Flow Batteries

Fuel Cells

A fuel cell is a device for converting chemical energy into electricity using an electrochemical reaction between a hydrogen-containing fuel and oxygen or suitable oxidizing agent. They differ from batteries in requiring a continuous input of hydrogen and oxygen to generate the energy. Given a constant supply of fuel and a way of disposing of the spent fuel, they can reliably produce electricity for extended periods. The difference between a fuel cell and a battery is that fuel cells need a continuous source of fuel and oxygen, usually from the air, to sustain the chemical reaction.

The six main fuel cell technologies under development are listed below. Each offers advantages for different applications.

- Alkali.
- Phosphoric acid.
- Solid oxide.
- Molten carbonate.
- Proton exchange membrane (PEM).
- Direct methanol.

Given the present enthusiasm for electric cars, PEM fuel cell development for cars is the most active area of development. Some technologies use high temperatures and unusual catalysts.

All types of fuel cells possess an anode, a cathode and an electrolyte that allows positively charged hydrogen ions to move across the fuel cell (see Fig. 11.1). A catalyst at the cathode accelerates fuel oxidation, generating hydrogen ions and electrons. The electrons are drawn from the anode to the cathode through an external circuit, producing direct current electricity. At the cathode, another catalyst causes hydrogen ions, electrons and

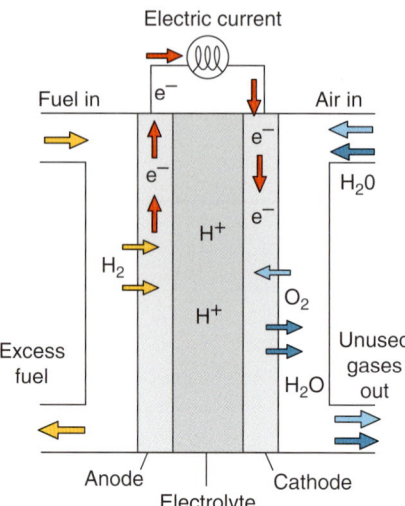

Fig. 11.1. Diagram of a proton-conducting solid oxide fuel cell. (From R. Dervisoglu, public domain via Wikimedia Commons.)

oxygen to react, forming water. In addition to electricity, fuel cells produce water, heat and, depending on the fuel source, very small amounts of nitrogen dioxide and other emissions. The energy efficiency of a fuel cell is generally between 40% and 60%; however, if waste heat is captured in a co-generation scheme, efficiencies up to 85% can be obtained.

Static fuel cell installed capacity

There has been a steady increase in the installation of fuel cells for electricity generation in homes and small businesses. Modular fuel cell systems are available to build small 5 MW to 63 MW power plants. Several large-scale fuel-cell power plants have been built in Connecticut, Delaware and California. It has been proposed to construct a 63.3-MW fuel cell power plant at Beacon Falls, Connecticut, USA. At present the largest power plant is a 59.9-MW fuel cell plant in South Korea. Japan and South Korea are leading proponents of fuel cell technology and South Korea presently plans to install 230 MW of capacity. London has a number of buildings with fuel cell systems: the Crown Estate in Regents Street and 20 Fenchurch Street (also known as the walkie-talkie building) have molten carbonate fuel cell systems, and Transport for London's Palestra building has a 200 kWe phosphoric acid fuel cell unit. Fuel cell systems for buildings are now available that offer combined heat and power. There is clearly a trend towards individual large buildings generating their own electricity. In Britain, the new Aberdeen Exhibition and Conference Centre will be equipped with a 1.4-MW fuel cell system, which will be one of the largest in Europe.

Environmental issues

The advantages of fuel cells in urban settings can be considerable because they silently generate electricity with little atmospheric pollution, in a compact area close to the point of power consumption. These advantages have led some to rather uncritically consider them to be environmentally advantageous over other electricity-generating methods. At present, there is simply too little operational experience of large-scale static fuel cell plants to reach a considered conclusion of the environmental balance sheet.

The key environmental issues revolve around the supply of the fuel. Hydrogen or methane need to be generated or extracted and there is a danger that all fuel cells will do in practice is to displace the pollution away from the point of consumption. They are only likely to be environmentally less damaging than other generation methods, such as nuclear power, if the hydrogen is generated using a truly renewable resource that does not cause large-scale environmental damage. Solar power and wind turbines are often suggested, but these technologies can have serious environmental effects, as discussed in Chapters 9 and 10 (this volume). Fuel cells may have a role in smoothing out the intermittent nature of solar and wind generation if hydrogen is generated during periods of excess generation using electrolysis.

Even if the hydrogen can be supplied in large quantities from a low environmental impact source, Tromp et al. (2003)[1] highlight the point that fuel cells will inevitably release hydrogen into the atmosphere and this could damage the ozone layer. They suggest that around 10–20% of the hydrogen would escape into the atmosphere. This problem would only become significant with large-scale use of fuel cells for power generation or transportation.

A key question that must be addressed when considering the environmental impacts of fuel cells is the source of the hydrogen. Dincer (2007)[2] estimated greenhouse gas emissions generated if natural gas, wind turbines, solar PV and gasoline were used as the energy source for hydrogen generation (Table 11.1). Taking into account environmental and economic criteria, he concluded that hydrogen production from wind energy via electrolysis is more consistent with sustainable development if greenhouse gas emission mitigation is the target, but traditional natural gas reforming

Table 11.1. Greenhouse gas emissions and air pollution during the life cycle of hydrogen and gasoline production and utilization (per MJ of lower heating value). (From Dincer, 2007.[2])

Source	Greenhouse gases (g)	CO (g)	NOx (g)	Volatile organic compounds (g)
Hydrogen from natural gas	84.8	0.0330	0.0761	0.0561
Hydrogen from wind energy	20.55	0.0142	0.0169	0.00132
Hydrogen from solar energy	30.6	0.0210	0.0251	0.00195
Gasoline from crude oil	84.0	0.876	0.115	0.170

is the more favourable route for air pollution reduction. It is notable that all hydrogen production pathways will generate greenhouse gases.

The recycling of fuel cells and the sourcing of exotic catalysts also have environmental consequences. Automotive fuel cells are designed for recycling and the presence of valuable materials aids the commercial viability of the processes.

Flow Batteries

A flow battery is a battery in which the ionic solution is stored outside of the cell and can be fed into the cell in order to continuously generate electricity. A schematic layout is shown in Fig. 11.2. One of the biggest advantages of flow batteries is that they can be almost instantly recharged by replacing the electrolyte liquid, while simultaneously recovering the spent material for regeneration. The commonest types of flow battery are redox, hybrid and membraneless.

Flow battery installed capacity

UniEnergy Technologies has installed a 2-MW, 8-MWh flow battery on the grid in Snohomish County in Washington state (see https://arstechnica. co.uk/information-technology/2017/04/washington-states-new-8-megawatt-hour-flow-battery-is-the-largest-of-its-kind/). This installation is currently the largest capacity containerized flow battery system in the

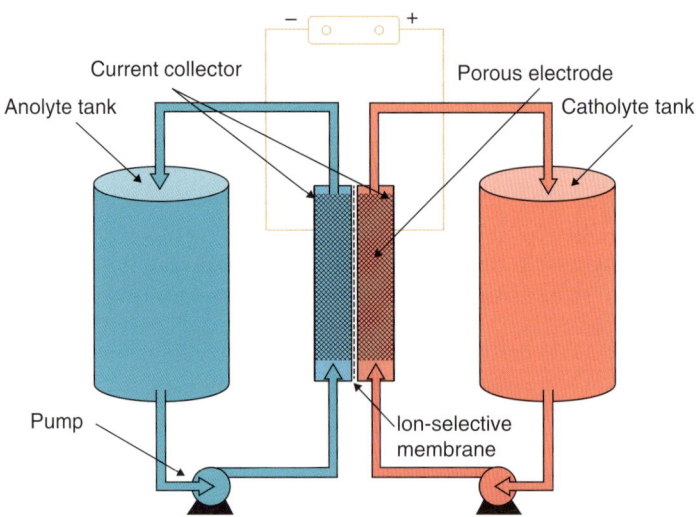

Fig. 11.2. A conventional redox flow battery illustration, showing two energy storage tanks and the electrochemical reaction cells. (From Colintheone, published under a CC BY-SA 4.0 license via Wikimedia Commons.)

world and is housed in 20 connected shipping containers. Flow batteries are most likely to be utilized in localities using renewable generation such as wind and solar, especially in locations without the benefit of a reliable grid. For example, vanadium redox batteries are installed at Huxley Hill wind farm in Australia and the Tomari Wind Hills in Hokkaidō, Japan.

Environmental issues

The greatest environmental issue relates to the ionic solutions, which may contain toxic compounds. It is possible to use non-toxic ionic solutions, but it is unclear if they can achieve the energy density of other designs. There is presently insufficient operational experience and too many competing designs to assess the environmental issues. A basic question that always needs to be asked is how is the ionic fluid manufactured and sourced?

Notes

[1] Tromp, T.K., Shia, R.L., Allen, M., Eiler, J.M. and Yung, Y.L. (2003) Potential environmental impact of a hydrogen economy on the stratosphere. *Science* 300(5626), 1740–1742.
[2] Dincer, I. (2007) Environmental and sustainability aspects of hydrogen and fuel cell systems. *International Journal of Energy Research* 31(1), 29–55.

12 Batteries

A battery is a device that converts chemical energy to electrical energy as required. Grid-connected large-scale batteries have long been desired by utility companies because they could aid the balancing of electrical supply and demand. In the last few years technological advances have made this possible. Batteries are not the only approach used for the storage of energy; another approach is pumped storage (see Chapter 2, this volume). A further approach is energy storage using compressed air; this is not considered further in this book because there are presently only two compressed air storage facilities, one in Germany and the other in Alabama, USA.

There are a number of battery technologies that have been used for utility-scale storage. The main types are:

- Lead–acid. The standard car battery we are familiar with is lead–acid. They are of declining interest in terms of large-scale electrical power storage, although large banks of lead–acid batteries are still used for emergency back-up on power plants.
- UltraBattery is a refinement of the lead–acid battery. A number of UltraBattery installations are sited at wind or solar farms in Australia. The UltraBattery was developed by CSIRO Energy Technology in Australia and is a hybrid energy storage device combining an asymmetric super-capacitor and a lead–acid battery. The capacitor enhances the power and lifespan of the lead–acid battery because it acts as a buffer during high-rate discharging and charging.
- Lithium-ion is a general term used to describe a range of lithium-anode-based rechargeable batteries. These are becoming the popular battery type for grid storage applications because of their relatively high energy densities, their lack of memory when only partially discharged, their ability to be charged and discharged over 1000 or more

Fig. 12.1. A 400,000 cell lithium-ion back-up power battery system installed at a San Diego Gas & Electricity's gas plant. (Photo courtesy of Yuan Gao, public domain via Wikimedia Commons.)

cycles and, probably of greatest importance, their falling cost. Li-ion batteries have three components: an anode, a cathode and a porous separator. The anode is composed of graphites with additives; the cathode is composed of layered transition metal oxides (e.g. lithium cobaltite (LiCoO) and lithium iron phosphates (LiFePO)).

- Flow batteries (discussed in Chapter 11, this volume).
- More exotic designs are at an assessment stage; for example, a molten-salt sodium–sulphur battery, which has been used in a 34-MW, 235-MWh facility at a Japan Wind Development project at Futamata in Aomori Prefecture, Japan.

Nickel–metal hydride batteries are of no importance for grid storage applications.

Installed Capacity

Battery storage is now being used to provide a replacement for peak demand generation in a number of countries, having been first introduced on a large scale in the USA, Japan, South Korea and Germany. Storage is particularly favoured in the USA because of the lack of a fully integrated grid. In the USA, utilities principally require storage to contribute to supply for about 4 hours of peak demand (typically between 4.00 pm and 8.00 pm). Recently, the improvement in both the energy-density of lithium-ion batteries and a reduction in their cost has made large-scale lithium-ion storage viable. In the USA alone, 1800 MW of new, mainly lithium-ion storage is expected to be online by 2021. Similarly, the National Infrastructure

Commission in the UK reported in 2016 that, if costs continue to fall as anticipated, up to 15,000 MWh of battery storage could be deployed in the UK by 2030. Battery storage is particularly attractive on islands with appreciable levels of intermittent renewable capacity. For example, the AES Kilroot Station in Northern Ireland plans a 50-MW lithium-ion facility (10 MW has already been installed). It is likely that as the amount of installed wind and solar capacity increases, the intermittent nature of the generation will increase the need for battery storage.

Environmental Impacts

The focus for utility-scale installations is on lithium-ion batteries, which are the dominant battery type. A 2013 EPA[1] study of Li-ion batteries for electric vehicles (https://www.epa.gov/sites/production/files/2014-01/documents/lithium_batteries_lca.pdf) concluded that batteries that use cathodes with nickel and cobalt, as well as solvent-based electrode processing, have the highest potential environmental impacts. These impacts include:

- resource depletion;
- global warming;
- ecological toxicity; and
- human toxicity.

Resource depletion, ecological toxicity and human toxicity are mainly linked to the release of cobalt, copper, nickel, thallium and silver. However, there is uncertainty about the impacts of aluminum and lithium because of the lack of appropriate toxicity data.[2] The largest environmental hazards are associated with the production, processing and use of cobalt and nickel–metal compounds, which may cause exposed people to suffer respiratory, pulmonary and neurological effects. The 2013 EPA report points out that nickel and cobalt also present significant environmental risks. Recycling of lithium-ion batteries is presently poor. There is general agreement for the need for stronger government action at the local, national and international levels to encourage recovery and recycling of lithium batteries.

There is a considerable literature on the ecological impact of lithium mining. Lithium is typically found in salt flats in areas where water is scarce, which is problematical as the mining process requires large amounts of water. The mining activity therefore risks contaminating and reducing local water supplies. Considerable areas of land are impacted by lithium extraction (see Fig. 12.2).

The European Commission in 2012[3] noted that lithium-ion battery production uses the equivalent of 1.6 kg of oil per kg of battery produced and releases up to 12.5 kg of CO_2 equivalent per kg of battery. However, it is estimated that per MegaJoule (MJ) of capacity, lithium-ion batteries are about half as toxic to humans as lead–acid batteries, and less toxic than nickel–cadmium batteries. Nickel–metal hydride and sodium–sulphur

Fig. 12.2. The Chemetall Foote Lithium Operation in Clayton Valley, a dry lake bed in Esmeralda County, Nevada, just east of Silver Peak, a tiny town that has been host to various kinds of mining for about 150 years. (Photo courtesy of Doc Searls, published under a CC BY 2.0 license via Wikimedia Commons.)

batteries are the least toxic to humans. The batteries that require the most energy for their production are nickel–metal hydride, at 90 MJ/kg of battery produced and lithium-ion, at 88–90 MJ/kg.

While lead–acid batteries benefit from their low cost and reasonable safety characteristics, they contain toxic materials and have low energy densities. About 47% of the total world lead production derives from recycled lead, the main source being lead–acid batteries. About 85% of all lead–acid batteries were recycled in the 1990s.[4]

McManus (2012)[5] noted that batteries were often an integral part of renewable low-carbon generating systems because they can store energy to meet variable demand. Further, batteries often comprise materials that have high environmental and energy impacts, some of which, such as lithium, are scarce natural resources. The result is that the overall impact of increasing our reliance on such sustainable or low-carbon systems may be an additional detrimental impact linked to battery use. Table 12.1 gives the manufacturing impacts per MJ capacity for six of the main types of battery. The analysis concluded that lithium-ion batteries have the most significant contribution to greenhouse gases and metal depletion. In terms of energy demand during manufacturing, nickel–metal hydride and lithium-ion batteries have a cumulative energy demand of 88–90 MJ/kg of battery weight compared with only 17 MJ/kg of battery weight for lead–acid batteries.

Table 12.1. Characterized data range for battery production per MJ capacity. (From McManus, 2012.)

Impact category	Unit	Lead–acid	Lithium-ion	Nickel–cadmium	Nickel–metal hydride	Sodium–sulphur
Climate change	kg CO_2 eq	5–7	17–27	10–15	16–20	2
Ozone depletion	kg CFC-11 eq	(2.24–3.35)E-07	(3.34–5.23)E-04	(7.64–12)E-07	(5.44–6.85)E-07	1.26E-07
Human toxicity	kg 1,4-DB eq	6–8	3–5	4–6	1.7–2.1	0.6
Photochemical oxidant formation	kg NMVOC	0.03–0.04	0.03–0.05	0.38–0.59	0.12–0.25	0.007
Particulate matter formation	kg PM10 eq	0.02–0.03	0.03–0.04	0.88–1.38	0.45–0.57	0.0085
Ionizing radiation	kg U235 eq	1.1–1.5	2.7–4.1	3.4–5.3	2.3–2.9	0.3
Agricultural land occupation	m^2	0.12–0.17	0.15–0.23	0.21–0.32	0.13–0.16	0.04
Urban land occupation	m^2	0.08–0.12	0.22–0.34	0.29–0.45	0.17–0.21	0.05
Natural land transformation	m^2	0.0011–0.0015	0.0021–0.0033	0.0019–0.0029	0.0013–0.0016	0.00033
Water depletion	m^3	0.07–0.1	0.121–0.191	0.25–0.39	0.14–0.174	0.024
Metal depletion	kg Fe eq	2–3	28–44	7–10	9–12	4
Fossil depletion	kg oil eq	1.8–2.6	2.2–3.4	3–5	5–6	0.5

Notes

[1] United States Environmental Protection Agency (2013) *Application of Life-Cycle Assessment to Nanoscale Technology: Lithium-ion Batteries for Electric Vehicles.* Available at: https://www.epa.gov/sites/production/files/2015-04/documents/lithium_batteries_abstract.pdf (accessed 8 December 2017).

[2] Kang, D.H.P., Chen, M. and Ogunseitan, O.A. (2013) Potential environmental and human health impacts of rechargeable lithium batteries in electronic waste. *Environmental Science & Technology* 47(10), 5495–5503.

[3] European Commission (2012) Environmental impacts of batteries for low carbon technologies compared. *Science for Environment Policy* 303. Available at: ec.europa.eu/environment/integration/research/newsalert/pdf/303na1_en.pdf (accessed 8 December 2017).

[4] Jolly, R. and Rhin, C. (1994) The recycling of lead-acid batteries: production of lead and polypropylene. *Resources, Conservation and Recycling* 10(1–2), 137–143.

[5] McManus, M.C. (2012) Environmental consequences of the use of batteries in low carbon systems: the impact of battery production. *Applied Energy* 93, 288–295.

13 Biofuels and Waste-Powered Generation

Biofuels

A biofuel is defined as a combustible material produced by recent biological production, such as photosynthesis or anaerobic digestion. In contrast, examples of non-biofuels are coal or mineral oil. Biomass is claimed to be the dominant source of renewable energy, contributing from 44% to 65% of total renewable output. In 2011, biomass-fuelled power generators provided around 10% of the world's energy supply.

Important biofuels include:

- Bioethanol: alcohol made by fermentation, mostly from carbohydrates produced in sugar or starch crops such as corn, sugarcane or sweet sorghum. These are mostly used as a petroleum substitute for vehicles.
- Biodiesel: produced from oils or fats using transesterification.
- Biogas: methane made from waste crop material through anaerobic digestion or bacteria.
- Biohydrogen: bacterial action can also produce hydrogen.
- Solid biofuel: wood chip, dried plants and other harvested organic material.
- Solid domestic and industrial waste.

For electrical power generation, solid biofuels are usually preferred. This is because solid fuel is easily transported in bulk and does not need specialized storage. Power stations do not require especially pure or consistent quality biofuel and can therefore use cheaper biomass sources. Woodchip is an important biofuel. Felled logs, fire-damaged wood, forest thinnings and sawmill residues are chipped for electricity generation in biomass-fuelled power plants. There are, however, many examples of power plants using vegetable oils and animal fats, including the ItalGreen plant in Italy, which is fuelled by palm oil and has a total output of 115 MW (see below).

It is common practice to substitute some coal with solid biomass in coal-fired power plants, although this may require modification of fuel systems and boilers. The co-firing of thermal plants is becoming common; by 2011, 45 GW of European capacity used some co-firing, while in North America about 10 GW of capacity is co-firing.

Installed Capacity

Global bio-power capacity increased by an estimated 6% in 2016, to 112 GW.[1] The USA is the largest producer of electricity from biofuels (68 TWh), followed by China (54 TWh), Germany (52 TWh), Brazil (51 TWh), Japan (38 TWh), India (30 TWh) and the UK (30 TWh). This is about 1.5% of total electrical generation. It should be noted that actual generation in the USA fell by 2% in 2016 from 2015 levels. In Europe, growth in electricity generation from both solid biomass and biogas is driven by the Renewable Energy Directive. Table 13.1 shows the relative contributions of the various biofuel types, using data for 2012.

General Ecological Effects of Biomass Combustion

By far the majority of biomass fuel is in the form of solid biomass which is burnt in a boiler to generate heat to run a steam turbine. Therefore, all of the environmental issues relating to the operation of steam turbines discussed in Chapter 5 may apply to biomass generation.

In addition, biomass generation has particular environmental issues linked to biofuels production and use. These include:

- the 'food versus fuel' debate on use of agricultural land;
- carbon emissions levels (is it carbon neutral?);
- sustainable biofuel production (will bio-cropping exhaust the soil?);
- deforestation and soil erosion (will forest be destroyed for bio-crops?);

Table 13.1. Global electricity generation using biomass fuel in 2012. (Data from *Renewables Global Status Report 2017*, http://www.ren21.net/status-of-renewables/global-status-report/)

Type	Installed capacity 2012 (GW)	Growth rate 2011 to 2012 (%)	Operating time (h)	Estimated electricity generation in 2012 (TWh)
Solid biomass	50	3.5	3500–7000	175–350
Biogas	14	10	3500–7000	47–95
Municipal solid waste	10	5	3500–7000	36–72
Liquid biofuels	2	20	3500–7000	6–12
Total biomass	76	5	3500–7000	265–529

- loss of biodiversity;
- impact on water resources;
- rural social exclusion and injustice; and
- nitrogen dioxide (NO_2) emissions.

These issues are discussed in more detail below under particular fuel types.

Intensive solid biomass production

Arguments in favour of growing crops of trees and other woody plants to burn as fuel for generating electricity are based on the renewable resource argument. It is reasoned that it is really a means of capturing solar energy via photosynthesis. It is therefore a renewable resource, which at least in the long run will not lead to a net increase in CO_2 emissions and therefore will not contribute to global warming. Essentially the CO_2 released by combustion is balanced by the CO_2 captured by photosynthesis and it is claimed to be 'carbon neutral'. There are, however, serious problems with this simple argument. The energy content of biomass is lower than with most fossil fuels and considerable areas are needed to cultivate the fuels. According to Metz *et al.*, (2001),[2] high-yielding short-rotation forest crops

Fig. 13.1. A biomass crop of willow, just across the road from Steven's Croft Wood Burning power plant. (Photo courtesy of M.J. Richardson, published under a CC BY-SA 2.0 license via Wikimedia Commons.)

or C4 plants (e.g. sugar cane or sorghum) can give stored energy equivalent of over 400 GJ/ha/year.

To look at the land use issue, pollution and energy balance in detail, we need to look at a specific example of a working generating station. The 10-MW ARBRE trial plant in Eggborough, South Yorkshire was the first wood-fuelled electricity-generating plant. It operated between 1998 and 2001, when it closed for financial reasons causing considerable economic harm to local farmers who had invested in the biomass production. The project involved a group of farmers growing and selling short rotation coppice (SRC) willow to a dedicated gasification plant under long-term contracts. The ARBRE plant was located alongside the pre-existing Eggborough coal-fired power plant and therefore close to pre-existing transmission infrastructure. The gasification plant employed atmospheric circulating fluidized bed technology, in which wood was burned to produce a gas which was passed to a gas turbine to produce electricity.[3] Using this gasification process improved the thermal efficiency of the plant.

The ARBRE trial plant generated 59.57 GWh/year, given an operating capacity of 85% and a thermal efficiency of 30.5%. The plant began operation in the fourth year of the project, when the first willow harvest was taken. It required 41,500 oven dried tonnes (odt) per annum of biomass. The land required for the project was 4.5 ha for the ARBRE gasification plant and about 2000 ha for the cultivation of SRC willow. This suggests that 1000 MW of installed capacity (the typical output of an average-sized coal-fired power station) would require 2×10^5 ha (1000 km^2) of agricultural land. This shows the unrealism of biomass agriculture in most countries, which can ill-afford to allocate such huge areas away from food production.

Cheng and Hammond (2017)[4] calculated total energy inputs and outputs over an assumed entire 19-year life-cycle of the ARBRE plant (Table 13.2). Note that considerable energy inputs are needed during cultivation, harvesting and transportation. Transportation is a notably high cost because of the low energy density of the fuel. Two important statistics are noted by Cheng and Hammond (2017):[4] first, the energy gain ratio

Table 13.2. A summary of life-cycle energy requirements and land-take associated with a 10-MW biomass gasification plant over its estimated 19 years of operation. (From Cheng and Hammond, 2017.[4])

Energy inputs	Energy input (kWh)	Energy output (kWh)	Land-take (km^2)
Cultivation	1,961,000 ± 725,000	–	20
Harvesting	1,150,000 ± 275,000	–	–
Transportation	13,310,000 ± 2,767,000	–	–
Chipping	11,930,000 ± 2,594,000	–	
Operation	–	893,600,000 ± 0	0.045
Waste disposal	–	–	–
Construction	54,720,000 ± 3,889,000	–	–
Decommissioning	547,200 ± 155,600	–	–
Totals	83,630,000 ± 10,400,000	893,600,000 ± 0	20.05

is about 10.85; and second, the energy density is the lowest of any of the technologies they considered, at 2.13 GWh/km^2. For comparison, they estimate that under UK conditions, which are not particularly favourable, solar PV (which is comparable in also capturing solar energy) has a poorer energy gain ratio of 3.6–6.2, depending on the technology, but a much enhanced energy density of 49–62 GWh/km^2.

The land use and energy yield calculations cannot be greatly changed using a different crop. Table 13.3 presents typical crop yields and energy content. Note that this table gives the upper yield of willow as 15 odt/ha/year. The ARBRE plant required 41,500 odt/year of willow from about 2000 ha, which requires a yield of 20.75 odt/ha/year. The ARBRE plant was working on higher than typical yields.

Figures typically quoted certainly indicate that the ARBRE plant at Eggborough did not have sufficient land under cultivation. Generally, a 10 MW power-generating plant with no time off-load will deliver 87,600 MWh (315,360 GJ) of electrical energy per year. Given an energy conversion efficiency of 35% or lower, the energy content of a year's supply of the biomass fuel must be at least 250 MWh or 900,000 GJ. Using an energy content of the biomass of 18 GJ/odt (Table 13.3), it will require 50,000 tons of biomass per year. With typical crop yields of about 10 odt/ha/year (Table 13.3), the 10-MW generating plant would need 5000 ha of good quality agricultural land allocated to biomass production. This land requirement is particularly ridiculous when compared with the 1 ha required for 10 MW of installed wind turbine capacity, although this would have a far lower availability.

Land management and maintenance of soil quality

It is important to consider whether intensive biomass production such as SCR willow can be maintained without loss of soil fertility and yield.

Potter (1990)[5] reported that up to 135.5 kg/year of nitrogen, 15.8 kg/year of phosphorus and 85.1 kg/year potassium were being removed by the stems of SRC crops producing over 10 odt/ha/year. He concluded that it is unlikely that this level of nutrient removal would limit SRC yield on a typical site in the short term. Such levels of deletion cannot be sustained long-term, so some fertilizer inputs would eventually be needed. There is surprisingly little discussion of this requirement.

Table 13.3. Typical crop yields and energy content of selected biofuels. (From UK Natural Environment Research Council (NERC), Towards a Sustainable Energy Economy (TSEC) programme.)

Crop	Crop yield (odt/ha/year)	Energy content (GJ/tonne)
Poplar	8–10	18.5
Willow	10–15	18.5
Miscanthus	10–13	17
Switch grass	9–10	18.3
Reed canary grass	5–15	16.2

One aspect that is not frequently discussed is the need for herbicides. As the Forestry Commission state,[6] 'it is not possible to overemphasize the importance of establishing an SRC crop in completely weed-free conditions'. It is clear that the Forestry Commission believe herbicide use is required. There may also be a requirement for fungicides and insecticides if the required yield is to be achieved. To some extent pest control can be achieved by growing a variety of genetic strains, but it seems unlikely that all growers will be able to avoid the use of chemical control agents.

Forestry and Agricultural Waste Product Solid Biomass

To date, much of the biomass fuel for European power plants has been imported from the Americas and Australasia. Such extended supply routes involving transoceanic shipping are inefficient, costly and inevitably have negative environmental impacts. While some of this is waste product from the forestry industry, there is also some wood felled for the purpose. These sources have become fully committed and some European plants have been unable to secure long-term supplies, resulting in plants being idle. Agricultural, household and industrial wastes can also be used as fuel. For example, in Australia and Latin America, sugar cane pulp is burned. Domestic and industrial waste generally has a lower energy content than wood chip. Waste-to-energy plants are discussed below.

Waste-to-Energy Plants

Waste-to-energy (WtE) describes a number of the processes used to generate electricity and/or heat from waste. Most WtE processes produce electricity and/or heat directly through combustion; some electricity is generated using a gas such as methane or methanol, or a liquid such as ethanol or synthetic fuel. In some cases, the primary purpose of the plant is the disposal of domestic waste and the electricity is a by-product.

Incineration is the most common WtE process (Fig. 13.2). Waste incineration has a poor public image because of poor practice in the past, with widely reported emissions of toxins such as dioxins and heavy metals as well as smoke and bad odours. Historically, incinerators emitted into the atmosphere appreciable quantities of:

- nitrogen oxides;
- sulphur dioxide;
- particulate matter;
- carbon monoxide;
- carbon dioxide;
- acid gases;
- lead;
- cadmium;
- mercury; and
- organic compounds such as dioxins and furans.

Fig. 13.2. Marchwood Energy Recovery Facility, a waste incineration plant near Southampton, England. It burns municipal waste and produces electricity. (Photo courtesy of Dr Richard Seaby, Pisces Conservation Ltd.)

All new WtE plants in OECD countries now must meet strict emission standards, including those on nitrogen oxides (NOx), sulphur dioxide (SO$_2$), heavy metals and dioxins. Some issues remain, including emission of fine particulates, heavy metals, trace dioxin and acid gases. There is also the need to dispose of toxic fly ash and incinerator bottom ash. Incinerators have electric efficiencies of only 14–28%.[7]

More recently a number of technologies that do not use direct combustion have been developed. Examples include:

- Gasification: produces combustible gas such as hydrogen or synthetic fuels.
- Thermal depolymerization: produces synthetic oil.
- Pyrolysis: produces combustible bio-oil.
- Plasma arc gasification or plasma gasification process: produces rich syngas.

A review of environmental concerns has been produced by The Blue Ridge Environmental Defense League.[8] Gasification and pyrolysis are similar processes; both decompose organic waste by exposing it to high temperatures with limited oxygen. Gasification occurs in the presence of a small amount of oxygen, pyrolysis without any oxygen present. Plasma arc gasification uses electrically generated plasma torches to convert waste material into gas and a slag by-product.

In addition to carbon monoxide and hydrogen, gasification also produces hydrocarbon oils, char and ash. Air emissions from gasification plants include:

- nitrogen oxides;
- sulphur dioxide;
- particulate matter;
- carbon monoxide;
- carbon dioxide;
- methane;
- hydrogen;
- chloride;
- hydrogen fluoride;
- ammonia;
- heavy metals mercury and cadmium; and
- dioxins and furans.

Pyrolysis units burning municipal solid waste are presently operating in Japan, Taiwan, the UK and Canada. One frequently voiced criticism of both incineration and gasification is that it disincentivizes waste reduction and directs waste away from composting and recycling.

The economics of gas transport require power generation units burning the gas to locate at or near gasification facilities, leading to a concentration of air pollution problems.

Liquid and Gaseous Biofuels

Liquid and gaseous biofuels are predominately used in transportation. Examples include methanol, ethanol, di-methyl esters, pyrolytic oil, Fischer-Tropsch gasoline and distillate, and biodiesel from vegetable oil crops (IPCC, 2001).[2] Brazil is the dominant producer of ethanol from sugar cane, while corn is used as the main feedstock in the USA. Biodiesel is produced from rapeseed in Europe. These products are all subsidized by government to some extent. Energy yields from different biomass sources can vary substantially. Vegetable oil crops have relatively low energy yields (40–80 GJ/ha/year) compared with crops grown for cellulose or starch/ sugar (200–300 GJ/ha/year).

As these liquid and gaseous fuels are mainly used as transport fuels, they are not considered in further detail here. However, there are some exceptions, such as the ItalGreen plant in Italy fuelled by palm oil, which has a total output of 115 MW. The ecological issues relating to palm oil production have been discussed extensively (see Fig 13.3 for a typical image of forest clearance for a plantation, which will inevitably result in a local loss of animal diversity). As Fitzherbert et al. (2008)[9] discuss, oil palm has replaced large areas of forest in South-east Asia. Oil palm plantations support much fewer species than forests, resulting in greatly reduced biodiversity.

Fig. 13.3. Deforestation in Malaysian Borneo showing original forest and oil palm plantations. (© NASA, public domain.)

Notes

[1] REN21 (2017) *Renewables Global Status Report 2017.* Available at: http://www.ren21. net/status-of-renewables/global-status-report/ (accessed 8 December 2017).

[2] Metz, B., Davidson, O., Swart, R. and Pan, J. (2001) *Climate Change 2001: Mitigation. Contribution of Working Group III to the Third Assessment Report of the Intergovernmental Panel on Climate Change.* Cambridge University Press, Cambridge, UK.

[3] Royal Commission on Environmental Pollution (RCEP) (2004) *Biomass as a Renewable Energy Source.* RCEP, London.

[4] Cheng, V.K. and Hammond, G.P. (2017) Life-cycle energy densities and land-take requirements of various power generators: A UK perspective. *Journal of the Energy Institute* 90, 201–213.

[5] Potter, C.J. (1990) *Coppiced Trees as Energy Crops.* Final report to ETSU for DTI on contract ETSU B 1078, Harwell, UK.

[6] Tubby, I. and Armstrong, A. (2002) *Establishment and Management of Short Rotation Coppice.* Forestry Commission, Edinburgh, UK. Available at: https://www.forestry.gov.uk/ pdf/fcpn7.pdf/$file/fcpn7.pdf (accessed 8 December 2017).

[7] The Blue Ridge Environmental Defense League (2009) *Waste Gasification: Impacts on the Environment and Public Health.* Available at: http://www.bredl.org/pdf/wastegasification. pdf (accessed 8 December 2017).

[8] Ibid.

[9] Fitzherbert, E.B., Struebig, M.J., Morel, A., Danielsen, F., Brühl, C.A., Donald, P.F. and Phalan, B. (2008) How will oil palm expansion affect biodiversity? *Trends in Ecology and Evolution* 23(10), 538–545.

14 Small-Scale and Mobile Electric Generators

There are a huge number of small-scale electric generators capable of generating up to 20 MW. Diesel generators are present in almost all large power plants and substations. They are used at most coal, gas and oil-fired power plants and almost all nuclear power plants as an emergency back-up power source to run safety critical pumps, fans, hydraulic units, battery chargers, turning gear motors for steam turbines, etc. They even act as back-up power supplies at hydroelectric generation facilities, giving emergency power to the spillway gates, for example. If an entire grid system fails, diesel generators are usually made available for what is termed in the USA a black start. Diesel, gas and petrol generators are also frequently used as back-up facilities for hospitals, airports and data handling facilities, and other buildings that need to maintain uninterrupted electrical power. Diesel generators dominate the back-up market, although some small-scale gas turbines are also used.

Diesel Generators

In many parts of the world, remote towns depend on diesel generators. Diesel generators are also used in small-scale green energy projects, for example, collecting methane from landfill sites. Oddly, in some parts of the developed world there has been a recent increase in small-scale generation. For example, in the last few years the UK has experienced an increase in diesel generation, referred to as 'carpark generators' or 'peakers'. These small-scale diesel generators provide a back-up power supply and have been feeding power into the grid when demand is high. This has come about as the UK government and the grid controllers work to maintain reliable supply because the reserve capacity at large plants has fallen to historically low levels. Small-scale diesel and gas generators account for the majority of new capacity brought forward under schemes to increase reserve capacity. Diesel generators have been successful in offering short-term capacity because they are relatively cheap when used for short periods. They are only likely to be displaced when battery and other power supplies are available in sufficient quantity to cover peak demands.

Environmental issues relating to diesel generators

There are good environmental reasons to limit small-scale generation to essential uses only. Exhaust gas emissions have many known adverse effects and these units suffer from the same problems as all diesel engines. Per unit of power generated, diesel generators produce greenhouse gas emissions at two to three times that of a combined-cycle gas generator. They also emit significant quantities of local pollutants such as nitrogen oxides and particulate matter, which can be extremely damaging to health. Within Europe, the Medium Combustion Plant Directive imposes tight emissions limits for smaller scale power stations such as diesel generators; however, it exempts from emission limits plants operating for less than 500 hours per year.

The present mix of generating capacity in some countries, including the UK, has resulted in a conflict between the need to ensure security of supply of electricity and the need to improve air quality. National Grid in the UK has recruited hospitals and other organizations to make back-up generators available at peak times and avoid blackouts. While this is desirable from a security of supply point of view, it is highly questionable from an air quality point of view because hospitals are located in urban locations and their generators often have exhaust systems that were not optimized to minimize atmospheric pollution but rather ensure continuity of supply.

Small-Scale Gas Turbines

Recent developments in micro-turbines technology use developments from military and aerospace industries that need lightweight, compact, high-powered generators and are not cost sensitive. More recently, manufacturers have noted the potential for turbines in small-scale power generation. Most micro-turbines are single-stage and shaft, low pressure ratio gas turbines. While in theory gas turbines can have a high efficiency, small-scale machines do not generally achieve the efficiency of reciprocating engines or large-scale power stations.[1] The size of micro-turbines varies considerably, often in the 25 to 80 kW range, although up to 1000 kW units are planned. They often use natural gas as fuel. Curiously, even very small gas turbines have been developed. The micro gas turbine developed by the Belgian PowerMEMS project has a rotor diameter of 20 mm and will produce a power output of about 1000 W.[2]

The efficiency and cost-effectiveness of gas turbines improves if the exhaust heat can be used. There are designs that use micro-turbines for combined heat and power. Turbine efficiency can be improved by increasing the operating temperature of the turbine, but this tends to increase NOx and CO pollution. Some micro-turbines are sold as having low-emission designs. For example, Capstone Microturbine Solutions sell machines that can operate on natural gas, propane, landfill gas, digester gas, diesel, aviation and kerosene fuels. The individual units are containerized. The 200 kW unit claims low NOx emissions of 0.14 g/bhp-h.

At present small-scale gas turbines contribute an insignificant amount of electrical power, but this may change. If so, there will be a need to consider the environmental implications in more detail.

Household-Scale Photovoltaic and Wind Generation

These are discussed in Chapters 9 and 10 (this volume).

Notes

[1] Pilavachi, P.A. (2002) Mini- and micro-gas turbines for combined heat and power. *Applied Thermal Engineering* 22(18), 2003–2014.
[2] PowerMEMS (2008) *Ultra micro gas turbine generator*. Available at: http://www.power-mems.be/gasturbine.html (accessed 8 December 2017).

15 Ecological Issues Relating to Transmission Lines

In most regions, large-scale electrical transmission is undertaken using cables supported on pylons. In urban areas transmission lines are placed underground or even along the beds of canals or rivers. However, placing high-voltage transmission lines underground is uncommon and can cost two to ten times more than an overhead line.

Impacts Linked to Above-Ground Transmission Lines

There are clearly aesthetic considerations and it is not uncommon for power lines in areas of outstanding beauty or cultural significance to be avoided or the lines taken underground. Aesthetic issues are not considered further here.

Impacts on birds

Birds are probably the animals most affected by above-ground transmission cables. Power lines are one of the most important causes of bird mortality. They kill birds following collision and through electrocution. Electrocution tends to occur when large birds, such as white-tailed eagles with a 2.45-m wing span, touch a cable when spreading their wings to take off from a perch on a pylon, causing a fatal short circuit. Such large birds may also, on occasion, touch two power lines simultaneously while in flight, causing electrocution. Further, transmission systems can also cause habitat loss, as some bird species avoid areas with power lines. Large birds such as raptors and storks are particularly vulnerable.[1] BirdLife International states that 'high losses (sometimes in excess of 500

Fig. 15.1. Power lines cut through forest along the bank of the Hudson River, New York. Indian Point Nuclear Power Plant is in the distance. (Photo courtesy of Dr Richard Seaby, Pisces Conservation Ltd.)

casualties per kilometre of power line per year) are reported from lines with multi-level arrangements, and with thin and low-hanging wires in sensitive areas'.

Table 15.1 summarizes impacts of power lines on different families of birds, as given in T-PVS/Inf (2003).[2]

Transmission line impacts are getting worse because of the need to build more transmission lines to support wind farms and solar installations (see Chapters 9 and 10, this volume). The American Bird Conservancy have identified what they claim are the ten worst sites for bird loss following collision with wind turbines and their associated power lines (https://abcbirds.org/10-worst-wind-energy-sites-for-birds/).

Guidelines are available to mitigate bird mortality,[4] which can be reduced with good design. However, it may not be possible to design overhead lines that are more visible to some species of birds because when inflight they look down rather than straight ahead. Studies reported by Martin and Shaw (2010)[5] on three particularly vulnerable species (kori bustards, *Aerdeotis kori*; blue cranes, *Anthropoides paradiseus*; and white storks, *Ciconia ciconia*) found that they typically look down while in flight. Therefore, the addition of tags, reflective markers, etc. to make power lines visible are ineffective.

Power lines can also lead to changes in species composition by changing the behaviour of birds. Coates *et al.* (2014)[6] studied ravens in US sagebrush

Table 15.1. Severity of impact on bird populations of mortality due to electrocution and collision with power lines for the different families of bird. 0 – no casualties reported or likely; I – casualties reported, but no apparent threat to the bird population; II – regionally or locally high casualties; but with no significant impact on the overall species population; III – casualties are a major mortality factor, threatening a species with extinction, regionally or on a larger scale. (Information from T-PVS/Inf, 2003.[3])

Taxonomic group	Electrocution impact	Collision impact
Loons (Gaviidae) and Grebes (Podicipedidae)	0	II
Shearwaters, Petrels (Procellariidae)	0	I–II
Bobbies, Gannets (Sulidae)	0	I–II
Pelicans (Pelicanidae)	I	II–III
Cormorants (Phalacrocoracidae)	I	II
Herons, Bitterns (Ardeidae)	I	II
Storks (Ciconidae)	III	III
Ibises (Threskiornithidae)	I	II
Flamingos (Phoenicopteridae)	0	II
Ducks, Geese, Swans, Mergansers (Anatidae)	0	II
Raptors (Accipitriformes and Falconiformes)	II–III	I–II
Partridges, Quails, Grouses (Galliformes)	0	II–III
Rails, Gallinules, Coots (Rallidae)	0	II–III
Cranes (Gruidae)	0	II–III
Bustards (Otidae)	0	III
Shorebirds/Waders (Charadriidae + Scolopacidae)	I	II–III
Skuas (Sterkorariidae) and Gulls (Laridae)	I	II
Terns (Sternidae)	0	I–II
Auks (Alcidae)	0	I
Sandgrouses (Pteroclididae)	0	II
Pigeons, Doves (Columbidae)	II	II
Cuckoos (Cuculidae)	0	II
Owls (Strigiformes)	I–II	II–III
Nightjars (Caprimulgidae) and Swifts (Apodidae)	0	II
Hoopoes (Upupidae) and Kingfishers (Alcedinidae)	I	II
Bee-eaters (Meropidae)	0–I	II
Rollers (Coraciidae) and Parrots (Psittadidae)	I	II
Woodpeckers (Picidae)	I	II
Ravens, Crows, Jays (Corvidae)	II–III	I–II
Medium-sized and small songbirds (Passeriformes)	I	II

habitat. The birds were building their nests on electricity power line poles and using the height afforded to target their prey. Raven numbers in the study area increased 11-fold between 1985 and 2009 and 58% of nests were located on transmission line poles. From their nests high above the sage brush, the ravens have perfect viewpoints and the height gives both a greater attack speed and an easier take-off. They are able to use these advantages to attack the nests of greater sage grouse and prey on other endangered species, including the San Clemente loggerhead shrike and the desert tortoise.

In some areas, the only acceptable approach may be to bury the cable; see the discussion of underground electric transmission lines below.

Effects linked to the electromagnetic fields

A magnetic field is created when electric current flows along a wire. According to various estimates, the upper limit to the magnetic field strength of an AC power transmission system varies from 10 to 50 µT. In the past, there was considerable concern about the long-term effects of magnetic fields linked to transmission lines on human health and animals. Some early studies suggested a link between transmission lines and diseases such as childhood leukaemia. No causal link has been proved and there is a growing belief within the scientific community that exposure to transmission line magnetic fields is not responsible for human ill-health. There are also persistent beliefs that transmission lines can alter animal behaviour. For example, some anglers favour fishing below power lines crossing rivers, believing the fish move into this zone. However, there is no supporting scientific analysis.

Implantable medical devices can suffer from electromagnetic interference. It is believed that the standard threshold, 1 gauss below which no effect occurs, makes any impact unlikely because this is 5 to 10 times higher than the electromagnetic field produced by high-voltage transmission cables.

Noise and light impacts

There may be noise and light disturbance during construction, but this is likely to be minor and short-lived. Noise will be generated by the construction equipment and vegetation cutting and logging. Transmission lines and equipment can generate an irritating humming noise often linked to the mounting of the conductor. Crackles or hissing noises may occur in high humidity or when foam from waves is blown onto the lines.

Transmission line systems need transformer substations and these can produce noise pollution. It is generally assumed that transformer noise pollution is a nuisance rather than an ecological impact.

Chronic noise exposure is now recognized as an important ecological issue. Barber *et al.* (2010)[7] point out that noise creates masking, the inhibition of sound perception. Birds, primates, cetaceans and rodents have all been observed to shift their vocalizations to reduce the masking.

Agricultural land

Transmission line pylons and other structures can cause the following agricultural impacts:

- Hinder the manoeuvring of machinery and not allow efficient patterns of work.
- Increase soil erosion.

- Create opportunities for weed and other pests to invade.
- Compact soils and damage drainage.
- Produce safety hazards, such as low-lying power cables.
- Hinder or prevent aerial spraying.

Wetlands

The construction and maintenance of transmission lines can damage wetlands in several ways, including:

- Heavy machinery can damage vegetation.
- Wetland soils, especially peaty soils, can be compacted.
- The construction of access roads can disrupt the natural drainage.
- Construction and maintenance activity can increase suspended sediment loads.
- Vehicles and construction equipment can introduce invasive plant species.

Transmission lines can be collision obstacles for waterfowl and large birds such as swans and geese (Fig. 15.2). In areas where bird collision risk is high, it is common practice to place markers on the top wire to make the lines more visible to birds. However, studies reported by Martin

Fig. 15.2. Power lines crossing Southampton Water, England. Such lines are a threat to large wetland birds. (Photo courtesy of Dr Richard Seaby, Pisces Conservation Ltd.)

and Shaw (2010)[8] indicate that some birds look down and therefore will not see the markers.

Woodlands

An electric transmission line right-of-way (ROW) is a strip of land used by electrical utilities to construct, operate, maintain and repair the transmission line structures. Building a transmission line through woodlands generally requires the clear felling of all trees and brush from the transmission path. The width of the cleared zone will vary with the size of the transmission lines and the voltage. For a 330-kV transmission line, the ROW width would typically be about 40 m. So a 1-km stretch of line results in the loss of 40,000 m^2 (4 ha) of forest.

Transmission construction impacts can include forest fragmentation causing damage to forest structure and the risk of biodiversity loss. The ROW creates an additional, very extensive, forest edge, allowing forest-edge plants and animals to invade the interior.

Underground Electric Transmission Lines

It is a common practice in residential areas to place low-voltage distribution lines underground. While this practice may reduce aesthetic and other impacts, it may increase others. For example, damage to tree roots resulting in tree death frequently occurs. High-voltage transmission lines differ from lower voltage lines in that above-ground structures are necessary to support the underground cable.

Underground transmission lines can have the following disadvantages:

- An increase in the area of environmental disturbance.
- The complete removal of small trees and brush along the transmission ROW.
- Increased construction and repair costs.
- Increased operation and maintenance costs.

They may also increase the costs of transmission if the lines need to be cooled.

Notes

[1] BirdLife International (2007) *Position Statement on Birds and Power Lines*. Available at: http://www.birdlife.org/eu/pdfs/Nature_Directives_material/BHDTF__Position_Power_Lines_and_birds_2007_05_10_.pdf (accessed 8 December 2017).
[2] Haas, D., Nipkow, M., Fiedler, G., Schneider, R., Haas, W. and Schuerenberg, B. (2005) Protecting birds from power lines: a practical guide to minimising the risks to birds from electricity transmission facilities. *Nature and Environment* 140.

[3] T-PVS/Inf (2003) Protecting birds from power lines: a practical guide to minimising the risks to birds from electricity transmission facilities. D. Haas, M. Nipkow, G. Fiedler, R. Schneider, W. Haas, B. Schürenberg, 2003 and published under Nature and environment, No. 140, Council of Europe Publishing, March 2005.

[4] Haas, D., Nipkow, M., Fiedler, G., Schneider, R., Haas, W. and Schuerenberg, B. (2005) Protecting birds from power lines: a practical guide to minimising the risks to birds from electricity transmission facilities. *Nature and Environment* 140.

[5] Martin, G.R. and Shaw, J.M. (2010) Bird collisions with power lines: failing to see the way ahead? *Biological Conservation* 143(11), 2695–2702.

[6] Coates, P.S., Howe, K.B., Casazza, M.L. and Delehanty, D.J. (2014) Landscape alterations influence differential habitat use of nesting buteos and ravens within sagebrush ecosystem: implications for transmission line development. *The Condor* 116(3), 341–356.

[7] Barber, J.R., Crooks, K.R. and Fristrup, K.M. (2010) The costs of chronic noise exposure for terrestrial organisms. *Trends in Ecology & Evolution* 25(3), 180–189.

[8] Martin, G.R. and Shaw, J.M. (2010) Bird collisions with power lines: failing to see the way ahead? *Biological Conservation* 143(11), 2695–2702.

16 Geothermal Generation

Geothermal electricity generation uses the heat found in the earth's crust. This heat in part dates back to the hot origins of our planet and also from heat generated by radioactive decay. While it is generally considered a renewable resource, it is possible to locally exhaust a heat source by removing more heat or water than is being generated by the earth. The actual design of the power plant depends on the temperature of the resource. Generally geothermal resources do not supply steam at the temperatures used in a conventional steam turbine. Flash plants use water at temperatures greater than 200°C; the steam is separated from the water, which is passed back into the reservoir. Binary-cycle power plants have been designed to utilize cooler geothermal reservoirs. Hot water is pumped from a geothermal well, passed through a heat exchanger and the cooled water is returned to the underground reservoir. A second fluid with a low boiling point, typically a butane or pentane hydrocarbon, is pumped at pressure through the heat exchanger, vaporized and then passed through a turbine. The vapour exiting the turbine is then condensed by cold air radiators or cold water and cycled back through the heat exchanger.

The first geothermal power station was built at Larderello, Italy in 1911. It was the only such plant until 1958, when another was built in New Zealand, followed in 1960 by the Geysers, California. Geothermal power is mostly developed in volcanically active parts of the world where tectonic plates meet. It is possible to use geothermal energy in geologically stable locations such as southern England, but in such areas it will never make a significant contribution to total electrical generation.

Fig. 16.1. Krafla Power Station, Iceland. Iceland's largest power station with
33 boreholes supplying geothermal energy for an installed capacity of 60 MW.
(Photo courtesy of Robin Somes, Pisces Conservation Ltd.)

Installed Capacity

The USA has the greatest installed capacity, although at about 3000 MW it
is of little significance (Table 16.1). To put the total installed capacity into
context, there is about 11,000 MW of geothermal capacity compared with
about 400,000 MW of solar.

Environmental Impacts

Geothermal water sources can be highly polluting. Water drawn from deep
under the ground can carry a mixture of gases including carbon dioxide,
hydrogen sulphide, methane and ammonia. If released to the atmosphere,
they contribute to global warming gases, may cause acid rain and can produce poor air quality with highly undesirable smells. Hydrogen sulphide
is both toxic and smells of rotten eggs. Geothermal water can also carry
toxic metals such as mercury and arsenic. One simple solution is to pass
the cooled water back into the reservoir.

Plant construction has been reported to have caused minor earthquakes because enhanced geothermal systems use hydraulic fracturing
(see p 120).

Table 16.1. Installed electric-generating capacity based on geothermal heat. (Data from Holm *et al.*, 2010.[1])

Country	Capacity (MW) 2010	Percentage of national electricity production	Percentage of global geothermal production
USA	3,086	0.3	29
Philippines	1,904	27	18
Indonesia	1,197	3.7	11
Mexico	958	3	9
Italy	843	1.5	8
New Zealand	628	10	6
Iceland	575	30	5
Japan	536	0.1	5
Iran	250		
El Salvador	204	25	
Kenya	167	11.2	
Costa Rica	166	14	
Nicaragua	88	10	
Russia	82		
Turkey	82		
Papua-New Guinea	56		
Guatemala	52		
Portugal	29		
China	24		
France	16		
Ethiopia	7.3		
Germany	6.6		
Austria	1.4		
Australia	1.1		
Thailand	0.3		
TOTAL	10,959.7		

Note

[1] Holm, A., Blodgett, L., Jennejohn, D. and Gawell, K. (2010) *Geothermal Energy: International Market Update.* Geothermal Energy Association, p. 7. Available at: http://www.geo-energy.org/pdf/reports/GEA_International_Market_Report_Final_May_2010.pdf (accessed 8 December 2017).

17 Minimizing Environmental Damage While Generating Electricity Cost-Effectively

What is Our Best Approach?

We desire clear, simple solutions and this can lead us to be attracted to those that offer what is claimed to be the single best approach. However, it is often the case that there is no optimal approach that covers all circumstances. As I will argue below, following consideration of the main environmental issues confronting our planet, this is the case for our choice of electricity-generating technologies. Further, the competing technologies and the associated control and distribution systems are undergoing rapid technological innovation. This pace of innovation is moving faster than our ability to fully assess their respective environmental merits. This is in fact the normal state of affairs: DDT, PCBs, CFCs, lead in petrol, non-biodegradable plastics and many other innovative products were introduced and seemed to offer great advantages to man. But, gradually, their environmental impacts were revealed and they were withdrawn from general use. The lesson is clear, be cautious and careful and do not overcommit to any technology until it has been thoroughly tested and monitored in typical working situations. Sadly, this is rarely the case for technologies linked to electricity generation. One feature I have noted is that those promoting different generating technologies will lock on to the dominant environmental concerns of the time and use it to sell their equipment. They will ignore, or hide, any environmental disadvantages, and governments and regulators driven by the need to address the dominant environmental issue will not always consider carefully the potential disadvantages. At the moment, this is the case with technologies that can make claim to being carbon neutral and therefore not adding to greenhouse gas emissions. Further, there is now an expressed aim in Europe to switch to electric cars: have the environmental consequences of the huge increase in electricity generation and the technologies to be used been thought through? I do not think so.

Imagine a situation in which we have perfected atomic fusion reactor technology and can now produce almost unlimited, cheap, electrical power from small, safe machines. There is no more need to combust coal, oil or gas and all cars are electric. The major causes of atmospheric and water pollution have been removed. The wind and solar farms and their impacts on animals and plants and the mining activities needed to support their manufacture have also vanished. With seemingly limitless power we can also afford to carefully recycle our waste, reducing the need for waste incinerators and we can also remove trace elements such as heavy metals from our air and water. This seems to be our ideal energy solution, but consider a little more deeply. The consumption of almost limitless electrical power would result in the unconstrained ejection of waste heat. Essentially man would be continuously introducing huge amounts of heat from our fusion reactors into our planet. We would struggle with thermal pollution. There would need to be constraints imposed on the use of energy. Although we can guarantee that a privileged elite class would not be so constrained, most of us would have to be. The conclusion is clear: all methods of electrical power generation are environmentally damaging if they are deployed without constraint. This also suggests that before we opt for what we consider to be the most environmentally favourable methods of generation and distribution, we should carefully consider if we are keeping our electrical consumption as low as is reasonably possible.

One of the most exciting developments over the last 30 years has been the growth in energy efficient technologies and the improved efficiency of household appliances. An appreciation of the engineering advances in energy efficiency recently achieved in household appliances such as freezers, ovens, dishwashers, washing machines and tumble dryers comes from estimates of the electricity saving that could still be obtained if everyone had modern appliances. For example, it is frequently claimed (e.g. see http://www.localgen.org.uk/microgeneration/understand-your-usage/) that if all European households replaced appliances more than 10 years old with new ones, 20 billion kWh of electricity would be saved annually, reducing CO_2 emissions by almost 18 billion kg. While the veracity of these figures is difficult to check, it is clear that appliances consume in the region of 33% of household electrical energy input and that modern appliances and lighting systems can reduce electricity consumption by 50% compared with older equipment. There is general agreement that before any increase in generating capacity is considered, it is worth considering the opportunities to reduce consumption via efficiency savings.

Data from the World Energy Council allows us to compare household electricity consumption in different countries and regions. The average consumption of electricity varies significantly between countries, depending primarily on the number of electrical appliances in the household. In 2016 it was about 1000 kWh/household in India, 2000 kWh in

Italy, 4000 kWh in Japan and 8000 kWh in North America.[1] In 2015, the average annual electricity consumption for a US residential utility customer was 10.8 kWh; in comparison, the UK figure is about 3.9 kWh and France about 5 kWh. I have spent extended periods in Italy, France, the UK and the USA and would view the overall quality of life to be lowest in the USA, which makes one wonder what they are actually gaining from their relatively huge electricity consumption.

The World Energy Council has also emphasized the key role of industrial improvements in efficiency. Industrial electric motors and electric motor-driven systems consume almost half of the total electricity generated and 70% of total industrial electricity. Motor average efficiency can be improved by 20–30%, which would reduce total global electricity demand by about 10–15% per year. This will likely be achieved in some parts of the world. The International Electrotechnical Commission has put in place an energy classification system for electric motors, which should make clear the relative efficiencies of different motors.

While domestic and industrial efficiency improvements are acting to reduce electricity consumption per person in the richest countries, there are two major trends acting to increase electricity demand. The first is the development of wealthier lifestyles in Africa, South America and Asia. It is inevitable that electricity consumption in India and China will increase greatly over the next 10 years. Second, there is now a growing movement towards the introduction of electric cars. Most electric vehicles consume about 10 kWh to travel 80–100 km. An extensive introduction of electric vehicles would seemingly require a major expansion of electrical generating capacity. We can get some idea of the required increase in capacity using some general calculations for the European Union presented on the Energy Matters website (http://euanmearns.com/how-much-more-electricity-do-we-need-to-go-to-100-electric-vehicles/). Table 17.1 shows that for the EU there would need to be an estimated 43% increase in installed generating capacity if all cars are electric. Such order of magnitude levels of increase would be typical for other parts of the world. Such numbers

Table 17.1. Approximate calculations for the required increase in installed electricity-generating capacity needed for full electric car use within the EU. (Data from the Energy Matters website - http://euanmearns.com/how-much-more-electricity-do-we-need-to-go-to-100-electric-vehicles/.)

Total number of cars	250,000,000
Average km driven/year	20,000
Total km driven/year	5,000,000,000,000
kWh consumed at 22 kWh/100 km	1,100,000,000,000
GWh consumed/year	1,100,000
2015 total electricity generation, GWh	3,231,000
% increase in 2015 generation	34%
Current installed capacity, GW	980
New capacity required at 30% load factor	419
% increase in capacity	43

do not seem inconceivable, and it is possible that a major push towards higher energy efficiency could reduce electrical demand by industry and in domestic consumption about 30% over the period when electrical cars are introduced, resulting in only a small increase in the installed capacity present in Europe and North America in 2015.

The energy policy debate over the last 20 years has been dominated by global warming. A strong consensus has been reached among scientists that CO_2 releases to the atmosphere are causing damaging increases in temperature and need to be reduced. However, even if recent warming caused primarily by increased CO_2 concentrations is accepted as occurring, it is legitimate to ask if it is the single most overriding important issue that needs to be addressed. In many western countries over the last 10 years, carbon-reduction has certainly been the key issue of focus when deciding on the technology of choice for electrical power generation. Before accepting the primary importance of CO_2 reduction, it is worth itemizing the major environmental issues we face. The following list is far from complete, but it gives an idea of the serious issues:

- Ozone layer depletion caused by chlorine and bromine found in chlorofluorocarbons (CFCs) and leading to increased UV radiation.
- Gaseous air pollution caused by the combustion of fossil fuels in vehicles, power stations and factories releasing SO_2, NOx and carbon monoxide. SO_2 is linked to acid rain.
- Dust pollution caused by mining, transport systems and other activities.
- Hydrocarbon air pollution caused by oil refineries and other chemical processes.
- Noise and light pollution affecting man and aerial, terrestrial and underwater life.
- Plastic and other waste product disposal.
- Water pollution caused by heavy metals, nitrates, phosphates, organic compounds, acid rain, urban run-off, PCBs, sex hormone disruptors, etc.
- Soil pollution and degradation caused by industrial waste, leaching from domestic waste tips, leaking oil stores, accumulation of pesticides and other agricultural chemicals, poor land management, etc.
- Nuclear contamination of land and water. This is a major issue in many parts of the world; for example, close to the Fukushima and Chernobyl power plants.
- Overpopulation increasing pressures on water, fuel and food resources and leading to over-exploitation. Our population has increased from 2,555,982,611 in 1950 to 7,481,000,000 in 2016.
- Depletion of fisheries, over hunting and other natural resource over-exploitation, including accidental increased mortality linked to human activities.
- Loss of biodiversity linked to human activities that cause direct harm to animals or plants and indirect harm via habitat damage.
- Deforestation and the increase in land under agriculture.
- Habitat loss and degradation caused by construction of roads, ports, cities, railways, airports, etc.

- Habitat fragmentation and the loss of migratory routes causing population decline.
- Introduction of non-native species and diseases causing loss of local species and ecosystem disruption.
- Global warming caused by the emission of greenhouse gases.
- Ocean acidification caused by CO_2 production.

We can now consider each of these issues in turn and consider if our choice of power-generating technology contributes to the problem.

Ozone layer depletion

The main agents of ozone depletion are chlorine- and bromine-containing anthropogenic chlorofluorocarbons (CFCs), halons, CH_3CCl_3 (methyl chloroform), CCl_4 (carbon tetrachloride), HCFCs (hydro-chlorofluorocarbons), hydrobromofluorocarbons and methyl bromide. Electrical appliances such as refrigerators and air conditioners used CFCs prior to 1995. The CFCs have been replaced by HCFCs, which are less harmful. As shown in Fig. 17.1, CFC concentrations are now declining, as is the total global depleting pollutants expressed as chlorine equivalents (EECl). However, HCFCs are increasing and although they deplete ozone to a much lower extent than CFCs, they will also eventually be replaced by hydrofluorocarbons (HFCs). However, HFCs are a greenhouse gas!

Essentially, the choice of electrical power-generating technology does not influence ozone depletion.

Gaseous air pollution caused by the combustion of fossil fuels

This is a particular problem linked to conventional large-scale power plants fuelled by coal, oil and gas. Small-scale diesel generators also cause gaseous air pollution.

Dust pollution caused by mining, transport systems and other activities

Dust pollution has always been a major problem for large coal-fired power plants. The mining, handling and transportation of the coal generates considerable dust pollution. Further, coal combustion results in the production of large volumes of fly ash, which is light and can be easily blown in the wind.

However, mining is also an essential activity for the manufacturing of components used in other power-generating technologies. For example, almost all power-generating systems use steel and concrete components to some extent and these are both derived from mining. Wind turbines need to be anchored in the ground and their towers are usually made of steel. Tidal and hydroelectric dams are massive concrete structures.

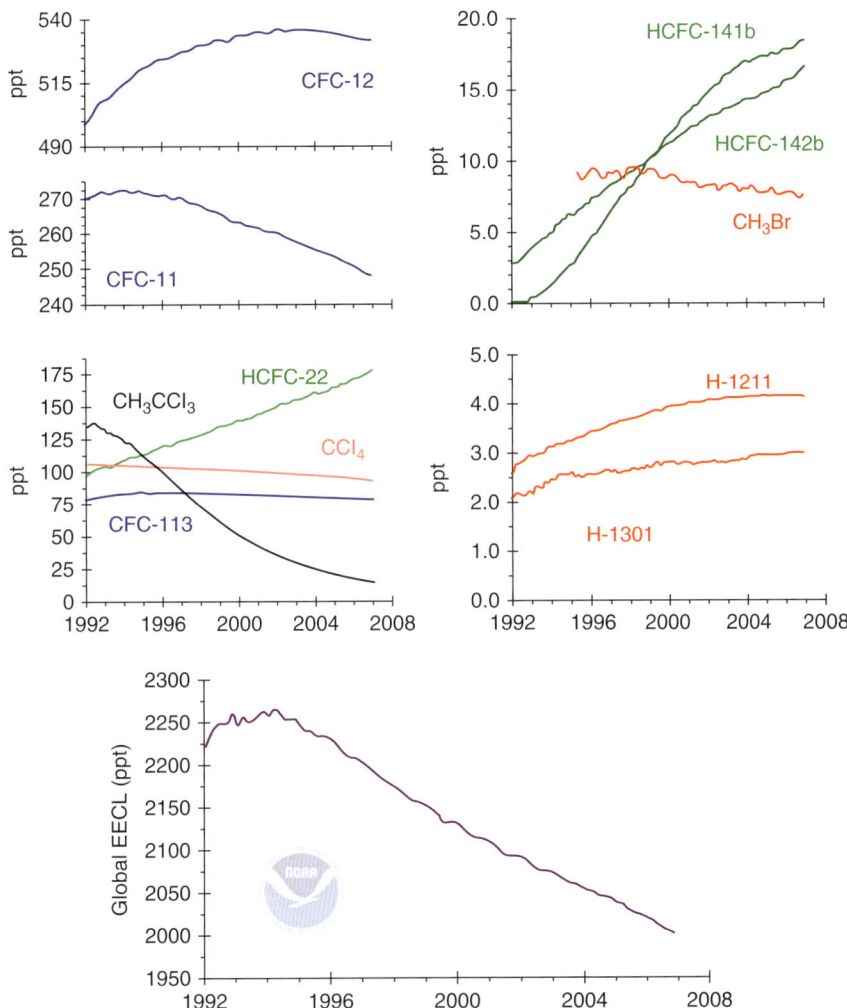

Fig. 17.1. Ozone-depleting gas trends and equivalent chlorine effect. Combined chlorine and bromine in the lower atmosphere or troposphere from the most abundant chlorinated and brominated chemicals controlled by the Montreal Protocol. Bromine is included as an ozone-depleting chemical because although it is not as abundant as chlorine, it is 45 to 60 times more effective per atom in destroying stratospheric ozone. Earlier measurements showed that the peak of equivalent chlorine (chlorine + 45 (or 60) times bromine) occurred at the surface between mid-1992 and mid-1994. The observed decrease is driven by a large and rapid decline in methyl chloroform and methyl bromide, gases that are regulated internationally by the Montreal Protocol. (Wiki commons data from https://commons.wikimedia.org/wiki/File:Ozone_cfc_trends.png.)

Photovoltaic cells require highly purified silicon. Mined silica is refined via a series of processes, including reacting carbon with the silica in an electric arc furnace, followed by a series of chemical processes to purify the silicon. Solar panels may also contain arsenic, bauxite, boron, cadmium, copper, gallium, indium, iron ore, molybdenum, lead, phosphate, selenium, tellurium and titanium dioxide. All are derived from mining.

In effect, the construction of all large-scale power-generating systems and their peripheral cables, transformers and pylons are based on mined products.

Hydrocarbon air pollution

This is a particular problem for large-scale oil- and gas-fired generation and small-scale diesel and petrol generators, and gas turbine systems using refined fuels.

Noise and light pollution

All forms of large-scale generation tend to cause some noise and light pollution. A case can be made for large-scale solar PVs causing almost no noise or light pollution when in operation, but even these systems will tend to have transformers that can generate local noise pollution.

There are both aesthetic and ecological issues relating to noise and light pollution. In many urban parts of the world light pollution makes it impossible to enjoy the wonder of the Milky Way or to see meteorites fall from the sky. Light pollution is a regular feature of power plants of all types, which often operate day and night. In addition, tall pylons are frequently fitted with navigation lights. Often little or no effort is made to control lighting to reduce the impact on night-flying insects or other animals.

Most forms of power generation create a noise nuisance. Hydroelectric and tidal generators cause underwater noise, as do offshore wind turbines. Large-scale conventional and nuclear power plants create considerable amounts of noise via their pumps, fans, cooling systems and coal mills. Transmission lines also frequently generate noise.

Noise pollution can produce ecological consequences. It is striking that the environmental consequences of noise pollution are often poorly understood. Are bats affected by wind turbine blade noise? Long *et al.* (2011)[2] argue that ultrasonic sound within the audial range of bats is generated, yet we know little about the effect on bats. In the case of offshore wind turbines, underwater noise may affect fish and diving mammals. Underwater noise pollution is also a significant issue in rivers and estuaries caused by shipping, large pumps and other machinery.

Plastic and other waste product disposal

This is an environmental problem to which electrical power generation of any kind contributes insignificantly. It can help to solve the problem by using waste plastic as a fuel. It is also true that the only positive feature I have noted for once-through cooling water systems is that the filter screens filter rubbish from the water together with fish and seaweed.

Water pollution

Large-scale conventional and nuclear power plants cause local water pollution via their heated cooling water discharges, which may also contain biocides such as chlorine. They also release trace amounts of heavy metals leached from their cooling water systems. Oil- and coal-fired plants release heavy metals and other contaminants with their flue gases. Mercury pollution linked to coal combustion is a major global issue. Hydroelectric, tidal, wind and solar PV generators cause little direct water pollution, although they can cause pollution during construction and decommissioning.

There are numerous examples of chemical pollution related to the electrical-generating industry. One example is polychlorinated biphenyls (PCB) pollution. PCBs were once used extensively in transformers and capacitors. The Hudson estuary, USA, offers a typical example of the problems that have occurred following their release to the environment. General Electric began using PCBs in its manufacturing processes at the Fort Edward and Hudson Falls plants in 1947 and 1952, respectively. The plants discharged manufacturing process wastewater containing PCBs directly to the Hudson River until 1977, when discharges to the environment were banned; residual contamination from both plant sites and sediments along the length of the river continue to release PCBs into the environment. The Fort Edward dam was removed in 1973, which resulted in the downstream release of an estimated 1,300,000 cubic yards of PCB-laden sediment. PCB contamination now exists in all 200 miles of river sediment downstream of Fort Edward.

Elevated levels of PCBs were first discovered in Hudson River biota in 1969, but their importance was not recognized for several years. In the early 1970s, the Department of Environmental Conservation began collecting limited data on PCBs in New York waters and fish. In 1973, the federal Food and Drug Administration adopted a 'tolerance' level for PCBs in food sold commercially, including fish, of 5 parts per million (ppm) in the edible portion, reduced to 2 ppm in 1984. At least 7 of the 11 species of Hudson River fish sampled between 1970 and 1972 had concentrations of PCBs that exceeded that level.

With the discovery of PCB contamination in fish came a decline in populations of top predators such as mink and river otter, potentially serious effects on fish, mammals, birds and other wildlife, and the necessity for measures to protect human health. In 1976 the first health advisory notice on the consumption of fish was issued; over the next few years a slew of restrictions and total bans were placed on commercial and recreational fishing and the sale of fish. These restrictions remain in place to date, to a greater or lesser degree, with a relaxation on recreational fishing allowing catch-and-release. Human consumption of fish and crabs was also regulated in a number of ways, ranging from advised maximum consumption to outright bans. Again, these restrictions mostly remain in force to date.

By the mid-1980s, PCB levels in fish from the Hudson had declined, although average levels in many species still exceeded the 2 ppm tolerance limit imposed in 1984. However, fish taken from the upper Hudson in 1992 and 1993 had PCB levels as high as those reported in the early 1980s. Additional releases from Allen Mill, at the Hudson Falls plant, may have contributed to the increased levels of PCBs detected in the fish. As a result, in 1994, advisories for the Troy Dam to Catskill reach of the lower river were changed from species-specific advice to advice to eat none for all species except American shad. PCB concentrations in Hudson River fish gradually returned to pre-1992 levels as the Allen Mill release was brought under control. Despite these declines, the fish remain contaminated with PCBs. The effects of the almost complete cessation of fishing on the species' populations is unquantified; however, it must have allowed a considerable increase in fish populations and consequent alterations in the food web.

Soil pollution and degradation

Almost all terrestrial power plants can have a local impact on the soil. Some of these impacts can be of considerable long-term importance. Old ash- or coal-handling areas in coal-fired power stations can create long-lasting contamination, as can oil storing and handling facilities. We do not, as yet, have the operational experience to determine if large-scale solar PV arrays placed on agricultural land produce long-term detrimental effects on soil quality.

Nuclear contamination of land and water

Serious releases of radionuclides have catastrophic consequences. The nuclear industry claims they are highly unlikely to occur. But they do occur and are not as unlikely as they are typically claimed to be by nuclear proponents. It should also be noted that terrorist actions are difficult to plan and design for. Early civilian nuclear reactor buildings were not designed to withstand a direct strike from a jet airliner. Nor were nuclear power plants designed to be resistant to worker sabotage. The rise of international terrorism brings into question many of the calculations concerning the low likelihood of radioactive contamination. The long half-lives of nuclear waste make the handling of nuclear material an onerous and costly operation, particularly in a politically unstable world.

Overpopulation

Population growth results in the need for ever more electric power generation. However, except to the extent that reliable electrical supplies increase life-expectancy, power plants do not contribute to this problem, which is probably far more important than many environmentalists are presently willing to accept. To some extent, the recent emphasis on climate change has taken the focus away from the very real problems linked to overpopulation – for example, the overuse of limited resources such as grazing land and fisheries.

Depletion of fisheries, over-hunting and resource depletion

While once-through cooling systems can damage local fish and crustacean populations in estuarine and enclosed water bodies, they do not generally lead to a detectable reduction in fisheries resources. This is usually because fishing takes a far greater proportion of the population than are killed by the power plants. Generally, power plants are not a primary area of concern when considering over-fishing and hunting.

There are situations where power generation is viewed as offering a positive effect because of the exclusion of fishermen and hunters. For example, offshore wind farms often act to exclude commercial fishing boats from sections of the sea bed. Further, the foundations and other underwater structures can offer reef-like environments that can be used by fish. Unfortunately, they may also protect some fish by reducing local bird densities.

It is also quite common for nuclear power plants and large conventional plants to have extensive fenced grounds, which can act as local nature reserves. Power plants situated on the coast often offer protected sites, free of human and dog interference, used by ground-nesting birds. Roofs and other tall structures are also used by gulls and other birds.

Loss of biodiversity and habitat degradation linked to human activities

All power-generating systems contribute to some extent to a loss in local biodiversity. It is clear that conventional and nuclear plants based on steam turbines and using once-through cooling will affect local organisms via impingement and entrainment losses, heated discharge effects and through the release of biocides such as chlorine, when used. Similarly, hydroelectric and tidal barrage schemes alter local aquatic conditions and will inevitably result in the loss of essential habitat for some species and their replacement by others. In recent years renewable generating systems that are considered low carbon, such as wind farms and large solar-generating systems, have portrayed their activities as environmentally positive because they reduce climate-altering gas emissions. However, this does not mean they do not have significant impacts on birds, bats, fish and aquatic life. The rush into low-carbon generation has meant that some of these impacts have not been fully understood and quantified. For example, the extent to which aquatic organisms flying between water bodies mistakenly land on solar PV panels is unknown. We still have a poor understanding of bird and bat losses at solar and wind turbine sites.

This issue is likely to grow in importance as large numbers of wind and solar farms are constructed.

The construction of power-generating systems inevitably requires mined materials. Mining results in habitat destruction and degradation. Power plants based on fuels such as coal, oil, gas and uranium inevitably have the greatest impact because of the extraction of natural resources. Given the low-quality coal frequently used, about 0.7 kg of coal is burnt per kWh of electricity generated, so a single 2000-MW coal-fired power station consumes $2 \times 10^6 \times 0.7 = 1.4 \times 10^6$ kg of coal per hour (or 1400 tonnes). Most of this coal comes from huge open-cast mines.

Electrical supply systems require the construction of overhead cables and pylons. These lines and pylons cause death and injury to large numbers of flying animals and can cause a change in the local balance between predators and prey. The burying of cables both on land and under the sea also has inevitable impacts linked to disturbance. Transmission line impacts are caused by all types of large-scale power generation, particularly solar and wind, which tend to be more widely distributed across the landscape.

The amount of land use required for different types of generation gives one measure of potential impact on wildlife. Fthenakis and Kim (2009)[4] presented a full life-cycle analysis of land use in electricity generation. A summary of their approximate results for different forms of generation is given in Table 17.2. Coal is particularly variable as the estimates depend on the type of mining, the quality of the coal, distance from the mine to the plant and other factors. They concluded, among the renewable options, the photovoltaic (PV) cycle requires the least amount of land and the biomass cycle the largest. Furthermore, in most situations, ground-mounted PV systems in areas of high insolation transform less land than coal plants when coal is supplied from open-cast mining. The PV-fuel cycle with an insolation of 2400 kWh/m²/year and 13% conversion efficiency will produce, on average, 40% more electricity than the coal-fuel cycle from the same area of land. Given the US average insolation of 1800 kWh/m²/year, solar PV will generate a comparable amount of electricity to the coal-fuel cycle per unit of land use. While nuclear shows the smallest land transformation, a nuclear plant is second to a biomass plant in the land actually directly occupied by the plant. In terms of land transformed, the biomass, wind farm and hydroelectric cycles are the most damaging.

Deforestation and the increase in land under agriculture

Conventional oil, gas and coal and nuclear power plants can contribute to deforestation via the mining and drilling for their fuel. Coal mining is particularly destructive. In the US Appalachian region, it is estimated that mining transforms in the region of 350 m² of land, often forested, per GWh of electricity generated[4].

Table 17.2. Approximate life-cycle land transformation for various fuel cycles based on a 30-year timeframe. (Based on estimates given by Fthenakis and Kim, 2009.[4])

Type of plant	Total land transformation including mining, transport and waste disposal (m^2/GWh)
Coal plant	250–900
Nuclear	125
Natural gas	300
Solar PV	250–300
Solar thermal	366–550
Wind farm	1,000–2,000
Hydroelectric reservoir	4,000
Biomass	12,600

Plants based on a biomass fuel cycle are particularly damaging in terms of land use and potential deforestation, and it is difficult to see any merit in power plants fuelled using palm oil and other tropical fuels which will compete for space with tropical forest. In the USA, the biomass fuel cycle has been estimated to require a staggering 12,600 m^2 of agricultural land per GWh of electrical generation. Further, the conversion of coal plants to burning wood chip, which is considered a carbon neutral, renewable, alternative to coal, has the potential to directly lead to the destruction of forest.

In both direct and indirect ways, electrical power generation often results in deforestation or forest degradation. Transmission line corridors are usually clear felled when passing through forest. It also frequently results in a loss of agricultural land. While onshore wind farms require considerable areas of land, it is the case that they can co-exist with agriculture and may offer a useful steady income to the land owner.

Habitat fragmentation and the loss of migratory routes

All large industrial developments contribute to habitat fragmentation. Conventional coal, oil and gas plants frequently cause habitat fragmentation via their fuel-supply systems. For example, coal is typically supplied by rail. Railways have been in existence for so long that we hardly notice the way they divide the land. An interesting example are the railway lines along the banks of the Hudson estuary, New York. In the 1840s the railroads were started on both banks of the Hudson. The building of the railways produced several effects:

- They acted as a barrier to stop the free movement of people and animals from the surrounding area to the river.
- The construction of railway bridges across the bays of the Hudson created large areas of slow-flowing water. This has developed over time into the reed-filled, marshy areas much in evidence today. This is, in effect, augmenting the changes caused by deforestation of the Hudson

watershed. The gravel littoral and sub-littoral habitats originally in many of these bays were lost to the aquatic community. Many species of crustacean live in the shallow waters of estuaries and these areas are often a habitat for young fish. The loss and disturbance of the shallow water environments in the river would have changed its characteristic fish fauna significantly, and with the appearance of new changed habitats, potentially allowed the proliferation of introduced species which otherwise would not have thrived.

- The clearance of vegetation for the railroad had a similar effect to the early forest clearances, though on a smaller scale, in increasing water and silt flux into the river.
- In more modern times, herbicide has been regularly used to clear railroad tracksides. This may affect local littoral or marsh habitats.

Oil and gas are frequently supplied in under- or above-ground pipelines and these require the existence of a disturbed corridor. Renewable generation can also lead to habitat fragmentation. Like conventional plants, it requires transmission lines and these frequently require the clearing of trees along a corridor, creating a barrier to the movement of some forest animals. Power lines can also interfere with the migration of birds.

Dams have a catastrophic effect on migratory activity. Again we can use the Hudson as an example. In 1825 the Troy Dam was built, stopping all migration of anadromous and catadromous fish beyond this point. The shad spawning once extended as far as Glens Falls. The full extent of the impact of this on the migratory fish species in the Hudson is impossible to quantify, since no reliable data exist from this period, but the loss of many miles of available spawning grounds for shad and other anadromous fish must have considerable effects on the populations. This dam and others built upstream of it have turned the Upper Hudson into a 40-mile chain of slow-flowing lakes. The native flowing water community in this region has presumably been replaced by one more adapted to the slow-flowing nature of the river.

Introduction of non-native species and diseases

There is a considerable literature on the introduction of warm water invasive species linked to warm water discharges at plants with direct-cooled steam turbine systems. This problem is probably of often minor consequence compared with the general introduction of aggressive invasive species brought about by human activity, particularly accidental land and water transportation. However, thermal pollution has long been recognized as an important issue because warm water can alter the local ecology. It is even the subject of an entire book entitled *Ecological Effects of Thermal Discharges* by T.E.L. Langford (1990)[5]. In some lakes in the warmer parts of the USA, thermal discharges have even caused the introduction of pathogenic organisms. For example, pathogenic *Acanthamoeba* and

Naegleria were isolated from cooling water discharges at several coal-fired power stations in the USA (Shapiro *et al.* 1980)[6].

Power plants can also offer habitats to invasive species. For example, the fouling of cooling water systems by the zebra mussel has led to some once-through cooling water systems using far more biocide than otherwise would have been required. North America provides a good example. In 1986, the zebra mussel, an inhabitant of fresh and brackish Eurasian waters, arrived in the Hudson estuary via the Great Lakes in the ballast water of ships. First seen in the Hudson at Catskill in May 1991, zebra mussels now inhabit the Mohawk River and the Hudson River from Albany to Haverstraw Bay. Within little more than a year of their arrival, the biomass of the mussels was greater than that of all other heterotrophic animals in the Hudson and reached an estimated 550 billion individuals, at an average density of 4000/m² over the freshwater tidal river. A secondary estimate was that, as filter feeders, the mussel population could filter the entire volume of the freshwater Hudson in 1 to 3 days. Their presence poses a number of very considerable threats to the ecosystem of the Hudson:

- Zebra mussels tend to colonize on rocky substrates in shoal areas, replacing or smothering any existing community that is in these habitats. Taxa of particular concern include Unionid and Sphaeriid clams. They also out-compete native mussel species for food and space, leading to a decline in native mussel populations.
- Phytoplankton and detritus are major food sources for lake and river food webs. Excessive removal of the phytoplankton by zebra mussels reduces the zooplankton species that feed on them and can result in fisheries-related impacts.
- Mussels can filter large amounts of water and reduce the available food in the water column. Their filtering activity increases water clarity and hence light penetration. This, too, can dramatically change the benthic community structure.
- Zebra mussels cause significant biofouling in water intakes. This requires higher levels of biocide to combat the problem and this could lead to secondary effects in relation to the biocide chemical being released into the environment.

Given their considerable numbers and their ecological effects (lakes and rivers colonized by the mussels often see 50–75% declines in phytoplankton and small zooplankton biomass, rise in water clarity of 50–100%, drop of more than 50% in filter-feeding zooplankton and native bivalves, and increase in macrophyte beds and animals associated with mussels), it is inevitable that their presence will have a profound effect on the food web of the Hudson. It is certainly the case that zebra mussels and other invasive bivalves have had an extraordinarily great impact on the ecosystems of the Great Lakes.

In conclusion, there is a risk that non-native species will be introduced while constructing and operating all types of power plant. Power-generating systems distributed over large areas, such as wind farms, risk disturbing and spreading pathogens and invasive species. However, movement along rail, ship and road corridors used by conventional plants are also considerable.

Global warming and ocean acidification

These environmental problems are particularly associated with methods of electrical power generation based on the combustion of carbon-based fuels. They are now considered such serious issues that in some countries, such as the United Kingdom, there is a drive to completely eliminate coal-fired plants. This is the key environmental concern where renewable generating technologies of all kinds are viewed as highly advantageous.

What is less obvious is that hydroelectric generation projects generate surprisingly large quantities of greenhouse gases (see p 17).

What is the Best Method of Generation?

My brief review in this chapter of the main environmental concerns and how the various methods of generation contribute to these problems shows that there is no obvious approach to electrical power generation that will minimize harm and provide the clear standout choice. The generating method of choice will, in part, be determined by the location and the relative vulnerabilities of the habitats present. If the region is home to a population of large aquatic birds, then wind farms may be disastrous. Conversely, if there are local natural gas supplies and it is possible to place a gas turbine on an industrial estate at the edge of town, this may be an ideal solution, particularly when combined with household solar PV and thermal panels. Local power plants have the great advantage that power losses in transmission are minimized and animal and plant losses linked to transmission lines are eliminated. There is also the possibility of increased efficiency from combined use of heat and power. It is particularly important not to simply view renewable technologies as intrinsically less damaging to our environment. They may reduce CO_2 emissions, but if this costs us the loss of the large raptors, or migratory fish, it will not be the best course to follow. Similarly, there seems no future in biomass production except on a small scale to utilize agricultural waste products. Wood chip is a waste product of the timber industry; however, the demand for wood chip is such that there is a danger that it will result in deforestation.

One direction of travel is clear: we need to use electricity as efficiently as possible and technological advances are making this possible. All methods of generation have environmental costs, so any reduction in the amount generated will reduce environmental harm. There is also considerable merit in improving the thermal efficiency of our houses and fitting them with solar PV and thermal panel water heating systems. While large-scale solar plants will create environmental damage to some extent, there is much environmental merit in small-scale PV generation, providing manufacturing and recycling can be undertaken efficiently. It is particularly important that we take a precautionary approach to recently developed renewable technologies. I cannot help feeling that the enthusiasm for wind turbines is a little like our previous enthusiasm for DDT. This miracle insecticide seemed to offer man huge advantages; the disadvantages only became apparent gradually as it passed up the food chains and the loss of top predators began to be noticed. Similarly, tidal and wave generators should be viewed as technologies with unknown downsides.

We are presently living in a period of rapidly changing and developing technologies for the generation of electricity. It seems unlikely that some technologies, such as nuclear and coal-fired plants, will continue to be widely developed or built. The huge 3200-MW Hinkley C power plant now under construction in England, at a cost of more than £20 billion with huge guaranteed subsidies, has all the hallmarks of a final grand gesture. Like the last few battleships, it is the final flowering of an irrelevant technology. However, power plants last for upwards of 50 years, so major changes in the pattern of power generation can only be observed over a generational time period. We will therefore all see a vast range of technological approaches in use for the foreseeable future.

Notes

[1] World Energy Council (2016) *World Energy Perspectives 2016.* Available at: https://www.worldenergy.org/wp-content/uploads/2016/10/Exec-Summary_EnergyEfficiency-A-straight-path-towards-energy-sustainability.pdf (accessed 8 December 2017).

[2] Long, C.V., Lepper, P.A., and Flint, J.A. (2011) Ultrasonic noise emissions from wind turbines: potential effects on bat species. *Proceedings of the Institute of Acoustics* 33, 907–913.

[3] Ibid.

[4] Fthenakis, V. and Kim, H.C. (2009) Land use and electricity generation: a life-cycle analysis. *Renewable and Sustainable Energy Reviews* 13(6), 1465–1474.

[5] Langford, T.E.L. (1990) *Ecological Effects of Thermal Discharges.* Elsevier Applied Science, London.

[6] Shapiro, M.A., Karol, M.H., Keleti, G., Sykora, J.L. and Martinez, A.J. (1983) The role of free-living amoebae occurring in heated effluents as causative agents of human disease. *Water Science & Technology* 15(10), 135–147.

Index

Note: Page numbers in **bold** type refer to **figures**
Page numbers in *italic* type refer to *tables*

Acanthamoeba 214–215
Acartia tonsa 80
accident 90, 92
acid mine drainage 97
acid rain 105–106, 200
acid seawater 110–113
acidification, ocean 206, 215
Aerdeotis kori 193
aesthetic issues 192, 197, 208
agriculture 157, 213
    impacts 195–196
    waste 185
air
    compressed 174
    emissions 187
    pollution 205
        diesel generators 190
        fracking 125–127
        gaseous 206
        gasification plants 187
        hydrocarbon 208
        hydrogen production 171, *171*, 172
        waste-to-energy plants 185
Albany Steam Generating Station
    (US) 68, *71–72*
algae 82
algal blooms 34
alkali/sulphide precipitation system 117
Alosa fallax 63, 64
aluminium oxide 99
ammonia 32, 115
Anguilla anguilla 63, 64, 136

Annapolis Royal Generating Station (US) 20
anode 169
Anthropoides paradiseus 193
appliances, household 203–204, 206
aquatic birds 160
aquatic ecosystem 16–17
    invasive species 215
ARBRE biopower plant (UK) 183, 184
Archimedes windmill 142
Ascophyllum nodosum 82
ash 99–101
    dump **100**
    lagoons **99**, 101
    ponds 100
Aswan High Dam (Egypt) 16, **16**, **17**
atomic fusion reactor 203
audiogram 146, **147**
avoidance behaviour 150

back-up power 189
Balakovo Nuclear Power Plant (Russia) **37**
Balbina Dam Hydroelectric Power Plant
    (Brazil) 11, **12**
barrier nets 73–74, *77*
barriers, behavioural 75
bass 63, 64
    hearing 150
    striped 72
bats 137, 140–141, 166, 167
    mortality 211
    sound impacts 208

batteries 174–175
    environmental impacts 176–179, *178*
    flow 172–173, **172**
    installed capacity 175–176
    lead-acid 174, 177, *178*
    lithium-ion 174, 175, 176, *178*
    nickel-cadmium 176, 177, *178*
    production 177–178
    recycling 176
    system **175**
Bay of Fundy (Canada) 21
behaviour
    avoidance 150
    birds 193
behavioural barriers 75
benthic life 82
benthic organisms 136
Bethlehem Power Plant (US) **44**
bioaccumulation 115, 116
biocides 209, 215
biodiesel 180, 187
biodiversity 136, 187, 205, 211–212
bioethanol 180, 187
biofouling 82–84, **85**
biofuels 180–181
    gaseous 187
    liquid 187
    solid 180
biological production 180
biomass 180
    importing 185
    land use 213
    production 182, **182**, 183, 184
biopower
    capacity 181, *181*
    energy requirements 183, *183*, 184
    land use 183–184, *183*, 212
birds 25, 32
    aquatic 160
    behaviour 193
    collisions 139, 140, 141, 166, 192,
        194, 196
    habitat loss 192
    migration 137
    mortality 139–140, 160, 165–166,
        192–194, *194*, 211
    predatory 38, 139, 140
    waterfowl 196
birds of prey 165, 192
bivalve animals 82, 83, 110, *114*, 215
black 161
black start 189
blowdown 43
blue cranes 193
borehole 93
bottom ash 101

Bowline Point Generating Station (US) 74
Brazil, Balbina Dam Hydroelectric Power
    Plant 11, **12**
breathing 106
brown-field sites 157
burial 93
burns 160, 165, **165**
butterflies 141

cables
    laying 32
    magnetic fields 31–32
    undersea 136
    see also transmission lines
calcium oxide 99
Canada, Bay of Fundy 21
carbon
    capture and storage 103–105, **105**
    scrubbing 105
carbon dioxide emissions 4, 102
    biomass production 182–183
    China 101
    coal- and oil-fired power plants
        101–105
    global 102, **104**
    hydroelectric generation 17–18
    reducing 205
    sequestration 103–105, **105**
    solar power 161
    United States 102
    wind turbines 131
carbon neutral resource 182
carbon neutral technology 202
care 202
carpark generators (peakers) 189
cars 202, 204
    generating capacity *204*
catalytic reduction 115, 116, **116**
cathode 169
caution 202, 217
cement 124
Chemetall Foote Lithium Operation
    (US) **177**
chemical energy 169, 174
chemical oxygen demand 108
chemicals, toxic 35
Chernobyl disaster 89, 92
China 88
    carbon dioxide emissions 101
    Jiangxia Tidal Power Station 21
    Longyangxia Dam Solar Park **159**
    solar power 154
    Three Gorges Dam 8, **9**, 12, **13**
    Yangtze River 13
chloride 109

chlorination 79, 82–84, **84**
chlorine, demand 83–84, **84**
chlorofluorocarbons (CFCs) 206
choices 1
Ciconia ciconia 193
classification system 204
cleanliness 2
climate change 102, 103, 139, 205, 206, 215
climates, dry 150
Clupea harengus 63, 134
coal 4
    combustion revenue 101
    consumption decline 7, 102, **103**
    dust 98, 99, 206
    mining 94–98, 212
    moving 97
    stock heaps 97, **98**
    sulphur content 106
coal-fired power plants 94–118
    biomass use 181
    carbon dioxide emissions 101–105
    fuel supply system 213–214
    greenhouse gas emissions 101–105,
        115–116
    layout **95**
    soil pollution 210
coal-handling system **98**
cobalt 176
cod 136
    hearing 150
collision 139, 140, 141, 166, 192,
        194, 196
colour 141, 160, 161
combustion 185
composting 187
concentrating solar thermal systems
        162–167
concrete 206
constraint 203
construction 35
    fracking wells 125, 126
    transmission cables 195
    wind farms 132–135
consumption 2
        and electrical appliances
            203–204
        future predictions **5**
        growth 6–7, 204
        and Human Development
            Index **4**
        United Kingdom 7
        United States **5**, 7, 204
containment failure 92
contamination 23
    radioactive 92
    water 92, 97, 98, 101, 109, 114–115, 126

control 2
conversion 2
coral reefs 34
cost
    decontamination 105, 107, 116, 117
    energy
        generation 1, 90, 92, 162
        storage 175
    nuclear generation 90
Crangon crangon 80
Crescent Dunes Solar Energy Project
        (US) **162**
Critical Thermal Maximum 80
crops 182–184, **182**, *184*
    vegetable oil 187
crustaceans, planktonic 80
cumulative impact 128

dams 8, **9**, 14, 16, **16**, **17**, 214
Dan River (US) 101
death rates, human 94
debris 65–66, 73, 74, 75
decontamination, costs 105, 107,
        116, 117
deforestation 187, **188**, 205,
        212–213, 216
demand, chlorine 83–84, **84**
deserts 157, 160
detritus 215
devices
    efficient 7, 203
    idling 2
    medical 195
Dicentrarchus labrax 63, 64
Didcot Power Station (Wales) **42**, 109,
        109, *110*
diesel generators 189–190
Dinorwig Power Station (Wales) 8, **10**
disasters
    Chernobyl 89, 92
    Fukushima Daiichi 89, 92, 98
disease 195, 214–216
diving mammals 150
dredging 133, 134, *135*
    fish injury 133, 134
drilling 98–99, 122
    gas 122, **123**, 124–128
dry climates 162
dry injection 107
dry lime scrubbing 107–108
dumping 100, 101
Dungeness Power Stations (UK) **78**
Dunkerque Power Station (France) 80
dust 97–98, 99, 206
dykes, porous 74

earthquakes 200
ecosystems
    aquatic 14
    disruption 206
    marine 22
EDF 90
education 3
eels 15, 63, 64
efficiency 101, 203, 205, 216
    device 7
    industrial 204
    thermal 216
Egypt
    Aswan High Dam 16, **16**, **17**
    Lake Nasser **17**
electric vehicles 202, 204
electrical fields 32
electricity
    advantages 2
    availability 1
    supply 174
electrocution 192, *194*
electrolyte 169
electromagnetic fields 31–32, 136, 195
Elm Road Generating Station (US) 117
emergency power 189
emissions
    air 187
    standards 186, 190
    see also carbon dioxide emissions
energy
    available sources 4–5, **4**
        comparisons 2
    chemical 169, 174
    costs
        generation 1, 90, 92, 162
        storage 175
    policy 205
    storage 162, 174–179
    thermal 162
entrainment **47**
    defining 45
    and dredging 134–135, *135*
    ocean thermal energy conversion
        plants 35
    in steam turbine cooling systems
        38, 43–56, *50*, *53*, **54**, **55**, *56*
    reduction devices 65–76, *68*, *69–70*,
        71–72, *73*, *77*
    see also fish
environmental impacts 1, 216
environmental issues 202, 205, 216
    nuclear generation 91–93, 205, 210
estuaries 25
estuary basin 23
ethanol 187
European Marine Energy Test Centre 28, **31**

eutrophication 109
evaporation 43
experts, industry 2
eyesight 167

fertilizers 184
Finland, Onkalo Nuclear Fuel Repository 93
fire 158
fish
    abundant 136
    aggregation devices 136
    baskets 66
    catches 17
    commercial 57, 63–64
    consuming 14, 117, 209
    delicate-bodied 65
    depletion 205, 211
    distribution 32, 82
    eggs 51, 72, 73, 133
    entrainment and impingement 43–56,
        46, *49*, **51**, *52*, *53*, **54**, **55**, *56*,
        134–135, *135*
        protection from 65–76, *68*,
            71–72, *73*, *77*
    escaping 134
    exhaustion 65, 66
    hearing 146, **147**, 149, *149*
    injury
        dredging impacts 133, 134
        hydroelectric generation 15
        noise impacts 149–150, **150**
        tidal barrage turbines 24, **24**, *25*
        tidal lagoons 27
    ladder **15**
    larvae 72, 134
    lift 16
    mercury contamination 117
    metabolism 80
    migration 14–15, 23, 57, 136, 214
    mortality 24, 25
        noise impacts 149
        pH impacts *114*
        reducing 65–76, *68*, *71–72*
        steam turbine cooling systems
            45, 46–56, *49*, *52*, *56*, 57,
            58–62, *68*, *71–72*
        temperature changes 80, 81, 82
        thermal discharges 80–82
        tidal barrage turbines 24, 25
    pH impacts 110, *111–112*, 113, *113*, *114*
    polychlorinated biphenyls
        contamination 209, 210
    predatory 38
    protection technologies 48, 54
    spawning 52, 133, 214
    stressed 80

suffocation 133
survival 45
    steam turbine cooling systems
        65–76, *67, 68, 69–70,*
        71–72
    temperature changes 80, 81, 82
    suspended sediments impacts 133–134
    swimming speeds 80
    temperature responses 80, 81, 82
fishing 136, 211
flash plants 199
flooding 11, 14
flow batteries **172**
    installed capacity 172–173
flue gas desulphurization 106–110, 117
fluid, working 32
fly ash *see* ash
food chain 109, 115
food web 115, 210
    aquatic 215
forest
    clearance 187, **188**, 205, 212–213
    fragmentation 197
    tropical 213
forestry waste 185
Fort Edward Dam (US) 209
fossil fuels 119
fracking 98–99, 119, 120, 122, 124–128,
        124, 200
    wells construction 125, 126
France 88
    Dunkerque Power Station 80
    La Rance Tidal Power Station 20, 25
Fucus vesiculosus 82
fuel
    cells 169
        environmental issues 170–171
        installed capacity 170
        solid oxide **170**
    reprocessing 92
    supply 169, 171, 213
    types **6**
        carbon-based 4
        energy yields 182, 183, *184,* 187
        fossil fuels 119
        land requirements 184, *184*
        nuclear 92, 93
    see also coal; gas; oil
fuel cells, recycling 172
Fukushima Daiichi nuclear disaster
        89, 92, 98

Gadus morhua 136
gas 4
    consumption growth 7
    dissolved 81

drilling 122, **123**, 124–128
    price 119
    production 120
    turbines 119, **121**, 190, 191
    well 122, 124–125, **126**, 128
gas-bubble disease 81
gas-fired power plants 119–128, **120**
    environmental issues 120–128, 213
    installed capacity 120
    technique 119
gasification 186
    air emissions 187
    plant 183
    plasma arc 186
gasoline production *171*
generators
    carpark (peakers) 189
    diesel 189–190
    noise impacts 208
    small-scale 189
geothermal generation 199
    installed capacity 200, *201*
gills 133, 134
glass 161
    panels 160
global warming 102, 103, 139, 205, 206, 215
Grain Power Station (UK) **120**
gravity foundations 133
greenhouse effect 102
greenhouse gases 12
    battery use 177, *178*
    coal-fired generation 101–105,
        115–116
    diesel generators 190
    emission standards 186
    fuel cells 170
    gas-powered generation 119
    gasoline production *171*
    geothermal generation 200
    hydroelectric generation 17–18, 216
    hydrogen generation 171, *171*
    see also carbon dioxide emissions
gypsum 108

habitat
    change 31
    creation 136
    damage 134
    destruction 11–12, 212
    fish 47
    fragmentation 206, 213
    littoral 214
    loss 26, 96, 192, 205
        birds 192
    marsh 214
harbours 145, 146

hazardous materials 158, 160–161, 176
    radioactive 88, 92, 98, 210
health impacts 3, 106, 127, 128, 161, 176,
        190, 195, 209
heat 162, 203
    death 80
    exchanger 88, 199
    exhaust 191
    recovery steam generator 119
    source 199
    transfer fluids 167
    waste 119, 203
heavy metals 109
herbicide 185, 214
herring 63, 134
    hearing *149*
High Marnham Power Station
        (UK) 85, **85**
Hinkley Point Nuclear Power Stations
        (UK) **40**, 64, 65, **89**, 217
horizontal slick-water fracking *see* fracking
hospitals 190
household appliances 203–204, 206
Hudson River (US) 209, 215
    estuary 213
Human Development Index (HDI) 3
    and electricity consumption **4**
human needs 1–7
humidity 44
humming 195
Hunterston Generating Station
        (Scotland) 82
hydrafrac 122
hydraulic fracturing 98–99, 119, 120, 122,
        124–128, **124**, 200
hydraulic pump 29
hydrochlorofluorocarbons 206
hydrodesulphurization 107
hydroelectric generation 5, 216
    carbon dioxide emissions 17–18
    and fish injury 15
    growth 9
    and habitat destruction 11–12
    installed capacity 9–11, *11*
    and land use 11–12
    life-cycle global warming
            emissions 17–18
    micro power stations 8
    technique 8
    water use 12–13
    wildlife impacts 14–17
hydrogen
    generating 171, *171*
    source 171
    supplying 169
hydrogen sulphide 200

ice-core data 116
Iceland, Krafla Power Station **200**
idling devices 2
illness 94
impact monitoring 159, 167
impingement
    defining 45
    ocean thermal energy conversion
            plants 35
    predicting 47–52
    in steam turbine cooling systems 38,
            43–65, *49*, *50*, **51**, *52*, *53*, **54**,
            55, *56*
    reduction devices 65–76, *68*,
            69–70, *71–72*, *73*, *77*
incineration 185
inconvenience 125
India
    Kurnool Solar Park 155
    solar power 155
Indian Point Nuclear Power Plant (US) 90
Indian River Power Station (US) 79
indigenous peoples 11
industrial efficiency 204
industrial sites 119
innovation 202, 217
insecticides 217
insects
    aquatic 162
    concentration 137
    death 166
    flying 161, 166
invertebrates 82
    planktonic 80
ionic solutions 173
islands 176
Itaipu hydroelectric dam (Brazil) **9**

Japan 89
    fuel cell technology 170
Jiangxia Tidal Power Station (China) 21

Kilroot Power Station (UK) **99**, **100**
Kingsnorth Power Station (UK) 64, 82, **95**
Kingston Fossil Plant (US) 101
kori bustards 193
Krafla Power Station (Iceland) **200**
Kurnool Solar Park (India) 155

lagoons, tidal 27
land
    degradation 96
    disturbance 96

use 11–12, 137, 139, 156–158, *213*
    biomass production 183, 212
    biopower 183–184, *183*, 212
    coal-powered generation 212
    nuclear generation 212
    solar power 212
    wind farms 184
landfill 100
larvae 72, 134
lead 177
lead-acid batteries 174, 177, *178*
lenses 162
life 1
    expectancy 210
lifestyles, wealthy 204
light 205, 208
    reflected 162
lime slurry 107, 108
limestone forced oxidation 107
lithium mining 176
lithium-ion batteries 174, 175
    environmental impacts 176, *178*
littoral habitats 214
living standards 2, 3, 204
Lochay Hydroelectric Power Station
    (Scotland) **10**
Longyangxia Dam Solar Park
    (China) **159**
louver screens 75, *77*
Lovett Generating Station (US) 74

magnetic field 32
Maine Yankee Power Station (US) 82
mammals
    diving 150
    marine 150
Marchwood Energy Recovery Facility
    (UK) **186**
marine ecosystems 22
marine mammals 150
marine organisms, pH impacts *111–112,*
    113, 114
marsh habitats 214
masking 195
medical devices 195
Medium Combustion Plant
    Directive 190
mega-traps 166
mercury 116–117
mesh 55, 57, 73–74
metals
    depleting 177
    heavy 109
    toxic 200
    trace 114–115

methane 12
    generating 171
    hydroelectric generation 18
methylmercury bioaccumulation 14
MeyGen project (UK) 25
micro-power stations 8
micro-turbines 190, 191
microfiltration 74, *77*
migration 14–15, 23, 32, 213–214
    birds 137
    fish 14–15, 23, 57, 136, 214
military needs 92
mine drainage, acid 97
miners 94
mining 92, 94–98, 206, 208
    coal 94–98, 212
    deep-shaft 96
    and habitat destruction 212
    lithium 176
    open-cast 96, **96**, **97**, 212
    silica 207
    strip 96
mirrors 162
modern technology 203
molluscs 113
    pH impacts *111–112, 114*
monarch butterflies 141
monitoring, impact 159, 167
Morone americanus 65
mortality
    bats 211
    birds 139–140, 160, 165–166,
        192–194, *194*, 211
    see also fish
motors 204
mussels 82, 83, 215

Naegleria 215
Nasser, Lake (Egypt) **17**
nature reserves 92, 211
nests 193, 194, 211
nets 73–74, *77*
nickel-cadmium batteries
    176, 177, *178*
nitrogen
    oxides 115–116
    removal 184
noise 26, 32, 97–98, 125, 205, 208
    background 146
    construction 195
    exposure 195
    fish injury 149–150, **150**
    generators 208
    solar power 208
    underwater 137, 142–151, 208

North Sea 63
Northern Ireland SeaGen project 26, **26**
Norway 9
nuclear disaster 89, 92, 98
nuclear fuel 92, 93
nuclear generation 4, 5, 6, 7
    cost 90
    environmental issues 91–93,
            205, 210
    installed capacity 88–90, *91*
    land use 212
    technique 88
nuclear reactor 88, **90**
    decommissioning 90, 93
    design **89**
nuclear waste 93
nutrient loading 34
nutrient removal 184

ocean acidification 206, 215
ocean thermal energy conversion
    closed-cycle 32
        plant diagram **34**
    ecological impacts 34–35
    installed capacity 34
    open-cycle 32
        plant diagram **33**
    plants 35
        diagram **33**, **34**
    technique 32
    viable ocean areas **33**
oil 4
    drilling 98–99
    palm 187, 213
    vegetable 187
oil palm plantations 187
oil-fired power plants 94–118
    carbon dioxide emissions 101–105
Oldbury Power Station (UK) 64
Onkalo Nuclear Fuel Repository
        (Finland) 93
oscillating water column 30, **30**
oscillating wave surge convertor 30, **30**
Out Newton Wind Farm (UK) **138**
over-hunting 205, 211
overpopulation 205, 210–221
overtopping device 30, **30**
oxidation, limestone forced 107
oxygen 81, 98, 108
    chemical 108
    dissolved 134
    supply 169
ozone layer
    damage 171
    depletion 205, 206, **207**

paint 141
palm oil 187, 213
particle motion 149
pathogenic organisms 214, 215
peak demands 189, 190
peakers 189
Pelamis Wave Energy Convertor 29, **30**, **31**
perforating 124, 125
pest control 185
pH impacts, on marine organisms
        111–112, 113, *113*, *114*
phosphorous removal 184
photosynthesis 182
photosynthetic organisms 79
photovoltaic cells 207
photovoltaic generation 154–162
    installed capacity 154–156, **155**, *156*
piling 132
    impact 142, **143**, **144**
    percussive **143**
    sound impacts 142–151
    vibratory 142, 145, **145**, **146**, **147**, **148**
planktonic invertebrates 80
planktonic organisms **47**, 79
plants, protected 158, 159
plasma arc gasification 186
plastic 208
plutonium 92
point absorber buoys 28, **30**
pollutants 32, 81, 84, *109*, *110*, 114–115
pollution
    air 205
        diesel generators 190
        fracking 125–127
        gaseous 206
        gasification plants 187
        hydrocarbon 208
        hydrogen production
                171, *171*, 172
        waste-to-energy plants 185
    coal 98, 101–117
    displaced 171
    dust 205, 206
    hydrogen production *171*
    light 205, 208
    noise 26, 32, 97–98, 195, 205, 208
    nuclear 92
    oil 98–99, *171*
    particulate matter 106
    soil 205, 210
    thermal 164, 203, 214
    water 205
        chemical 209
        coal dust 98
        coal-fired power plants 98, 109,
                114–115

eutrophication 109
flue gas desulfurization 109
fly ash 101
fracking 126, 127
mining 97
nuclear generation 92, 210
polychlorinated biphenyls 209
contamination 209, 210
ponds 101
poorer populations 1
poppers 75
porpoise 151
potassium removal 184
Potomac River Generating Station
(US) 65
Potomac River (US) 52
pout 136
power plants (stations)
local 216
micro- 8
oil-fired 94–118
see also coal-fired power plants;
gas-fired power plants
precipitation, alkali/sulphide system 117
predation 134
predators 209
pressure groups 1
pressurized water reactor 88, **89**, **90**
prestige 1
primary production 23
protected species 158, 159
proton exchange membrane fuel cell 169
pulverized fuel ash 99–101
pumped storage plants 8
pumps, variable-speed 75, 76
pylons 212
pyrolysis 186, 187

radar 141
radiation, thermal 103
radioactive material 88, 92, 98, 210
storage 93
waste 93
radon 127
railways 213
rain, acid 105–106, 200
La Rance Tidal Power Station
(France) 20, 25
raptors 139, 140, 165, 192
ravens 193, 194
recharging 172, 174
recycling 187
batteries 176
fuel cells 172
wind turbines 160–161

reefs
artificial 136, 211
coral 34
regulations
emissions 186, 190
fish consumption 209
fly ash 101
renewables 5, 31, 211, 212, 213, 214, 216
batteries 177
biofuels 180, 182
fuel cells 171
geothermal 199
growth 7
hydropower 11
wind power 141
reserve capacity 189, 190
reservoir 8
residential areas 197
resources
depletion 176, 211
exploitation 205
extraction 212
respiratory disease 115
respiratory impacts 106
RidgeBlade Wind Turbine 142
Ristroph screens 65–70
rivers, downstream flow rates 14
Rosteon Generating Station (US) 67
Royal Society for the Protection of Birds
(RSPB) 139
Russia, Balakovo Nuclear Power Plant **37**

salinity 68
salmon 149
hearing *149*
scavenging 82
Scotland
Hunterston Generating Station 82
Lochay Hydroelectric Power
Station **10**
screens
cylindrical wedgewire 70–72, *77*
fine mesh 73, *77*
louver 75, *77*
Ristroph 65–70, *69–70*, *71–72*, *77*
travelling 65–70, 73
water intake 65–73, *68*, *69–70*, *71–72*,
73
sea level 76
seabed structure 136
SeaGen project (Northern Ireland) 26, **26**
seawater 32
acid 110–113
scrubbing 110–115
security 190

sediments, suspended 133–134
selective catalytic reduction 115,
        116, **116**
Severn, River (UK) 14–15
        estuary 64, 133
        tidal barrage 20, **22**
sewage 81
shading 157
shale rock formations 122
shipping 136, 146
Sihwa Lake Tidal Power Station (South
        Korea) 20
silicon 161, 207
        dioxide 99
Sizewell B Nuclear Power Station
        (UK) **90**
slurry, lime 107, 108
smell 200
smelting 105
soil
        pollution 210
        quality 184–185, 205, 210
solar energy 182
solar flux 166
solar panel array **155**, **158**
        small-scale 161
solar power 154–167
        batteries 174
        biodiversity impacts 211
        India 155
        installed capacity 162, **164**
        land use 212
        and mining 208
        noise pollution 208
        small-scale 217
solar thermal systems, concentrating
        162–167
sole 63
        hearing 150
Solea solea 63
Solucar PS10 (Spain) **163**
solutions 202
        ideal 203
sound 75, 140
        bats 208
        duration 149
        perception 195
        piling 142–151
        pressure 142, **144**, 145, **146**
        underwater 142–151
South Korea
        fuel cell technology 170
        Sihwa Lake Tidal Power Station 20
Spain
        solar power 162
        Solucar PS10 **163**

species
        non-native 206, 214–216
        protected 158, 159
        richness 136
        vulnerability 192, 193
spoil heaps 96
steam 199
steam turbine cooling systems
        biodiversity impacts 211
        biofouling control 82–84, 215
        closed 37, 43–44, 119
        combined wet and dry 44–45
        condenser 36
                air-cooled 43
        cooler, air-cooled fluid 43
        cooling tower 38, 41, **42**, 44
                biofouling 85, **85**
                blowdown 84–85
        cooling water discharge **40**
        direct 37–38, **39**, **40**
        dry air-cooled 37, 43–44, **44**
        entrainment 38, 43–56, *50*, *53*, **54**,
                55, *56*
                protection technologies 65–76,
                        68, *69–70*, *71–72*, *73*, *77*
        environmental issues 38, 43, 44,
                56–65
        evaporative 40–43, 44, *77*
                wet-cooled 37
        and fish
                mortality 45, 46–56, *49*, *52*, *56*,
                        57, *58–62*, *68*, *71–72*
                survival 65–76, *67*, *68*, *69–70*,
                        71–72
        functioning 36
        impingement 38, 43–65, *49*, *50*, **51**,
                52, *53*, **54**, **55**, *56*
                protection technologies
                        65–76, *68*, *69–70*, *71–72*,
                        73, *77*
        indirect 37
        once-through 37–38, **39**, **40**, **45**,
                56–65, *77*, 88
        open 37, 44–45
                recirculating 40–43, **41**
        thermal discharges 76–82
        water intake 38, 45–48, 51, 52, 54–57,
                64–76, 83
                screens 65–73, *68*, *69–70*, *71–72*,
                        73, *77*
steel 206
storks 192
storm events 65–66
streamers 165, 166
submerged pressure differential 30, **30**
subsidence, mine 96

suffocation 133
sulphur
    dioxide 105–115
        removing 106–110
    oxide 106
summer 79, 82
surface attenuators 29–30, **30**
suspended sediments
    concentration 133
    impacts on fish 133–134
swimbladder 150

technology
    carbon neutral 202
    fish protection 48, 54
    modern 203
Tellina tenuis 82
Temora longicornis 80
temperature 68, 76, 78, **79**, **84**, 108, 199
    benthic life response 82
    changes and fish
        mortality/survival 80, 81, 82
    differential 32
    water 68, 76, 78, **79**, 80–82, **84**, 108
terrorism 90, 93, 210
Texas Barnett shale 122, **123**
thermal discharges 76–82, **79**
    and fish mortality 80–82
thermal energy 162
thermal impacts 34, 80–84
thorium 98
Three Gorges Dam (China) 8, **9**, 12, **13**
tidal barrage 20–21
    animal injury 23, 24
    bird impacts 25
    cross-section **22**
    La Rance (France) 20, 25
    River Severn (UK) 20, **22**, **23**
    turbines
        design **23**
        and fish injury/mortality 24,
           24, *25*
tidal generation 19
Tidal Lagoon Swansea Bay (Wales) 27
tidal lagoons 27
tidal power **21**
tidal range 19, 21
tidal stream 19, 77
    environmental issues 26–27
    generators 25–27
tidal turbine 26
time 217
Topaz Solar Farm (US) 156–157,
    157, **158**
toxic chemicals 35

toxicity 176
transmission lines 2
    above-ground 192–197,
        193, **196**
    biodiversity impacts 212
    bird impacts 192–194, *194*
    right-of-way 197
    underground 197
    see also cables
transportation 105, 183, 187, 206
Trisopterus luscus 136
tritium 92
turbines
    gas 119, **121**, 190, 191
    micro-turbines 190, 191
    tidal 26
    tidal barrage 23–25
    see also steam turbines; wind
        turbines
twaite shad 63, 64

UltraBattery 174
undersea cables 136
undersea equipment 31
United Kingdom (UK) **99**, **100**
    ARBRE biopower plant 183, 184
    battery storage 175–176
    coal-powered generation **103**
    electricity consumption 7
    energy
        consumption 7, 204
        security 190
    fuel cell technology 170
    Marchwood Energy Recovery
        Facility **186**
    marine energy resource **21**
    Out Newton Wind Farm **138**
    power stations
        Didcot **42**, 109, *109*, *110*
        Dinorwig 8, **10**
        Dungeness **78**
        Grain **120**
        High Marnham 85, **85**
        Hinkley Point Nuclear **40**, 64,
           65, **89**, 217
        Hunterston 82
        Kingsnorth 64, 82, **95**
        Lochay Hydroelectric **10**
        Oldbury 64
        Sizewell B Nuclear **90**
    River Severn 14–15, 20, **22**,
        64, 133
    Tidal Lagoon Swansea Bay 27
    tidal power **21**
    wave power **29**

United States of America (USA) 4, 5
  battery storage 175
  carbon dioxide emissions 102
  Chemetall Foote Lithium
        Operation **177**
  Crescent Dunes Solar Energy
        Project **162**
  Dan River 101
  electricity generation **6**
  energy consumption **5**, 7, 204
  flow batteries 172
  Fort Edward Dam 209
  Hudson River 209, 215
      estuary 213
  Kingston Fossil Plant 101
  Potomac River 52, 65
  power (generating) stations
      Albany Steam 68, *71–72*
      Annapolis Royal Generating
          Station 20
      Bethlehem **44**
      Bowline Point 74
      Elm Road 117
      Indian Point Nuclear 90
      Indian River 79
      Lovett 74
      Maine Yankee 82
      Potomac Rivor 65
      Rosteon 67
      William States Lee III Nuclear 84
  solar power 155
  Topaz Solar Farm 156–157, **157**, **158**
uranium 88, 92, 98
urea 115
Urothoe brevicornis 82

vaporization 166, 199
vegetable oil crops 187
vegetation clearance 214
vehicles, electric 202, 204
velocity caps 74
visibility 193, 196
vulnerabilities 216

Wales
  Dinorwig Power Station 8, **10**
  Tidal Lagoon Swansea Bay 27
warbler **165**
waste 2, 93
  agricultural 185
  disposal 100, 185, 208
  forestry 185
  hazardous 100
  nuclear 93
  products 185
  radioactive 93
  reduction 187
waste-to-energy plants 185–187
water
  air-saturated 81
  contamination 92, 97, 98, 101, 109,
        114–115, 126
    lithium 176
    nuclear 205
  cooling 36–38, 44, 54–57, 76–82, 108,
        113, 209
  cooling tower blowdown 83–84
  deep 145
  density 78
  discharges 34, 76–82, **79**
    chlorine concentration 83–84
    invasive species impacts 214
    pH changes 110–113, *111–112*,
          113, *114*
    pollutant concentration 109,
          109, 209
    wet lime scrubbing 108–109
  drinking 127
  estuarine 52, 53, 57
  filtering 215
  flow 76
  flowback 127
  freshwater 52
  groundwater 127
  hot 199
  intake 38, 45–48, 51, 52, 54–57,
        64–76
    biofouling 215
  location 47, 51, 53
  metal-rich 97
  pollution 205
    chemical 209
    coal dust 98
    coal-fired power plants 98, 109,
          114–115
    eutrophication 109
    flue gas desulfurization 109
    fly ash 101
    fracking 126, 127
    mining 97
    nuclear generation 92, 210
  pressure **143**
  release 101
  salinity 68
  seawater 32, 110–115
  shallow 214
  slow-flowing 213
  sources 36, 47, 52
  stored 127
  surface 97, 127

temperature 68, 76, 78, **79**, 80–82,
        84, 108
use 12–13, 44, 158, 164
volume 47, 51
warm 214
waste 127
waterfowl 196
wave power 31–32
    generation types 28–31
    United Kingdom **29**
Wellman Lord process 115
wells, gas 122, 124–125, **126**, 128
wet lime scrubbing 108–110
wetlands impacts 196–197
Weyburn-Midale Carbon Dioxide
        Project 104
white perch 65
white storks 193
wildlife
    conservation organizations 139
    criterion 117
    hydroelectric generation impacts
        14–17
William States Lee III Nuclear Generating
        Station (US) 84
willow 183, 184
wind
    speed 140
    turbines 129–151, **130**

decommissioning 141
disadvantages 217
installed capacity 129, **131**, **132**
and mining 208
noise impacts 208
recycling 141–142, 160–161
wind farms
    batteries 174
    biodiversity impacts 211
    cables 32
    construction 132–135
    land use 184
    offshore 131–137
    onshore 137–141, **138**
windmills 129
winter 79
wood 5, 183
woodchip 180, 216
woodlands impacts 197
world regions 2, 3
    energy production **3**
    fuel types 4–5

Yangtze River (China) 13

zebra mussel 215
zooplankton 215